Control Systems

Continuous and Discrete

Control Systems

Continuous and Discrete

Victor J. Bucek

PRENTICE HALL, Englewood Cliffs, New Jersey 07632

Library of Congress Cataloging-in-Publication Data

Bucek, Victor J.
   Control systems.

   Includes index.
   1. Control theory. I. Title.
QA402.3.B782 1989   629.8'312   88-32434
ISBN 0-13-171752-9

Editorial/production supervision and
   interior design: **Kathryn Pavelec**
Cover design: **Wanda Lubelska Design**
Manufacturing buyer: **Robert Anderson**

If your diskette is defective or damaged in transit, return it directly to Prentice-Hall at the address below for a no-charge replacement within 90 days of the date of purchase. Mail the defective diskette together with your name and address.

   Prentice-Hall, Inc.
   Attention: Ryan Colby
   College Operations
   Englewood Cliffs, NJ 07632

LIMITS OF LIABILITY AND DISCLAIMER OF WARRANTY:
The author and publisher of this book have used their best efforts in preparing this book and software. These efforts include the development, research, and testing of the theories and programs to determine their effectiveness. The author and publisher make no warranty of any kind, expressed or implied, with regard to these progams or the documentation contained in this book. The author and publisher shall not be liable in any event for incidental or consequential damages in connection with, or arising out of, the furnishing, performance, or use of these programs.

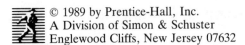
Printed in the United States of America

10 9 8 7 6 5 4 3 2 1

ISBN   0-13-171752-9

Prentice-Hall International (UK) Limited, *London*
Prentice-Hall of Australia Pty. Limited, *Sydney*
Prentice-Hall Canada Inc., *Toronto*
Prentice-Hall Hispanoamericana, S.A., *Mexico*
Prentice-Hall of India Private Limited, *New Delhi*
Prentice-Hall of Japan, Inc., *Toyko*
Simon & Schuster Asia Pte. Ltd., *Singapore*
Editora Prentice-Hall do Brasil, Ltda., *Rio de Janeiro*

. . . to our children's children

Contents

2 Laplace Transforms   *9*

3 Transfer Functions   *37*

4 Stability   *49*

5 Block Representation of Control Systems   *60*

Part II   Discrete

10   Digital Control Systems   *180*

11   The *z*-Transform   *195*

12   Discrete Transfer Functions   *217*

13   Discrete Control System Analysis   229

14   Discrete Proportional Integral Derivative (P.I.D.) Control   247

15   Discrete Compensator Design   266

16   Discrete State-Variable Analysis   284

Preface

The material in this text is now being used in a two-semester course in the final year of a Control Systems Technology program. The first semester deals with continuous control; the second, with discrete control. The text is also suitable as a reference for both the practicing professional in the controls area and the university student about to undertake his or her first control systems course.

The text is in two parts: In Part I we begin with Laplace transforms and, using the frequency domain, we study the implications of controlling a second-order system. The popular Proportional plus Integral plus Derivative (P.I.D.) control scheme is studied in detail. System stability is predicted with the Routh table, root-locus, and Bode plots. The popular compensating networks are covered, including lead, lag, and lead-lag. We leave the classical methods and investigate the time domain utilizing state variables. This modern approach to control theory is introduced, and the ideas of controllability and observability are covered.

In Part II we turn our attention to a computer closing the loop. We introduce z-transforms and the idea of sampling in our discrete domain. Once the discrete compensating network is generated, we will provide the computer algorithms to perform the function in terms of difference equations. The discrete PID control algorithm is presented in the BASIC, C, and Pascal languages. System stability is studied using the root locus on the z-plane. Digital filters are presented and the frequency response of discrete functions. The sampling theorem is used as a base for the sampling time required for our control system. We look at the discrete state variables and the idea of controllability and observability.

The two parts of the text are closely linked in terms of examples given and material covered. The first part looks at continuous control, the second at discrete control. This linkage encourages the reader to understand the implications of both domains and provides the flexibility to travel from one to the other. The examples used for theory clarification are dimensionless and lack lengthy process definition. Their purpose is simply to amplify the basics and not to obscure them with colorful applications. The problems at the end of each chapter reflect the real world in terms of quantities to be controlled and utilize the basics learned in the chapter.

In order to appreciate the derivations of the mathematics required in the control systems environment, the reader should be familiar with integral and differential calculus. For modern control theory, matrix algebra is also required. The reader should be aware of the various transducers available and the quantities measured. In the case of microprocessors, the reader should be fluent in at least one of the many available on the market.

This text presents basic principles of control systems and remains timeless and hardware independent. At first glance the text may appear very mathematical—and in fact it is. The basics of control systems technology are firmly rooted in mathematics, and the text treats mathematics as a means of understanding these basics. The reader will emerge with a firm grasp of the basics, ready to apply his knowledge to practical problems. The text encourages thinking, as the control systems technologist of tomorrow will have to be creative. As for practicality, that comes with experience in hands-on environments. For this reason, the text covers the material mathematically and refrains from citing large numbers of so-called practical examples, which in many cases create an illusion of experience without offering any real substance.

ABOUT THE SOFTWARE DISK

The disk contains several BASIC programs to enhance the various aspects of analog and digital control. It contains programs to plot system response, P.I.D. simulation, and P.I.D. root locus in the continuous sense. In terms of digital control, programs include various responses of digital filters, aliasing, and root locus of a second-order system in the discrete sense. The programs are easy to use and follow the text material.

Use of these programs will help teachers present the text material to the students.   It will also provide the students a mechanism for learning by experimentation.

The software can be modified, and the author welcomes constructive criticism.

ACKNOWLEDGMENTS

In any technical field it is the fundamental principles that withstand the test of time, and it is the writer's duty to recreate the same through the eyes of existing technology. I am indebted to all those, past and present, who have developed the fundamental principles; to my parents who in their own way provided me with the inspiration to see; to my wife and children for their patience and understanding; to Colin Bogle for his efforts on the many illustrations; to my students for their encouragement; to the proofreaders for their constructive criticism; and to the staff at Prentice Hall for putting it all together.

VICTOR J. BUCEK

Control Systems

Continuous and Discrete

Part I

Continuous

1

Introduction to Continuous Control Systems

1.1 DEFINITION OF A CONTINUOUS CONTROL SYSTEM

In this age of automation and product development it is interesting to note the constant reference to the "control system" that is required or being used. The utterance of this term awakens many different ideas as to its meaning. Is this some technical sorcery practiced by a handful of engineers—or is it something more tangible?

Simply put, a **control system** is an arrangement of physical components that promote self-regulation or the regulation of another system. Figure 1.1 provides a block diagram of the components involved and the relevant terms associated with a typical control system. The system requires an input known as the **setpoint**. This provides the system with a goal for self-regulation. To attain this goal the system requires knowledge of performance. This is obtained at the **summing junction** as an **error**, which should approach zero. To obtain this error we require **feedback** of the system **output**, which is performed by a suitable **transducer**. If this error is to diminish, we require an inversion of the feedback, or what is more commonly known as **negative feedback**. This error is then passed to the **compensation network** for amplification or further processing and becomes the **control action**. This signal becomes the driving force

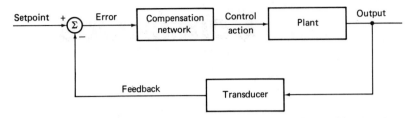

Figure 1.1 Components of a control system

for our **plant** under control. The plant responds by modifying the output, and the control system is complete.

If the compensation network generates a continuous control action, then the control system becomes a **continuous control system**. This will not be the case when we use a computer to generate the control action. To understand a simple control system, let's take a realistic example and perform the control process.

EXAMPLE 1.1

Consider a tank that is 15 meters in height, initially empty, and is to be filled to a height of 10 meters. The tank has no liquid leaving and has a proportional valve at the source to facilitate the filling process. A level transducer is available that provides a 1-volt output for every 1 meter of liquid in the tank. The components of the control system and the characteristics of the plant are provided in Figs. 1.2 and 1.3, respectively.

From Fig. 1.2 it is evident that our plant is comprised of the tank and proportional valve. The transducer in this case converts liquid level to a corresponding voltage output. This output provides the required feedback, and the summing junction adds the setpoint to the inverted feedback (negative feedback) to create the error. The error is multiplied by unity (compensation network) and becomes the control action (control voltage) for the

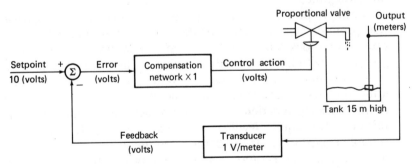

Figure 1.2 Control system for Example 1.1

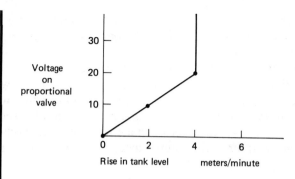

Figure 1.3 Input-output characteristics for the plant

proportional valve. This signal will open the valve and begin to fill the tank with liquid. (It is interesting to note that in our control system the *controlled variable* is liquid level, yet the *control medium* is voltage. This fact illustrates the importance of transducers in the field of controls technology.) To obtain 10 meters of liquid we require a setpoint of 10 volts at time $t = 0$ minutes, and Table 1.1 provides all the corresponding changes in the parameters at 1-minute intervals. Would you consider this to be a continuous form of control?

From Table 1.1 it is evident that our control system is trying to attain the level of 10 meters indicated by our setpoint. The system is accomplishing this feat by diminishing the error to zero and in doing so will eventually close the proportional valve. Our control action in this case is equal to the error of the system, and it should be clear that if we increased our gain in our compensation network, our system would diminish the error more quickly. This would require that our compensation network provide a means

TABLE 1.1 Parameter Changes in Example 1.1

Time (minutes)	Setpoint (volts)	Feedback (volts)	Error (volts)	Control action (volts)
0	10	0.00	10.00	10.00
1	10	2.00	8.00	8.00
2	10	3.60	6.40	6.40
3	10	4.88	5.12	5.12
4	10	5.90	4.10	4.10
5	10	6.72	3.28	3.28
6	10	7.38	2.62	2.62
7	10	7.90	2.10	2.10
8	10	8.32	1.68	1.68
9	10	8.66	1.34	1.34
10	10	8.93	1.07	1.07

for the designer to change the speed of response of the control system. Upon this component of the control system we will spend a considerable amount of time in the chapters to come.

    If we plot the setpoint and plant output as a function of time and also plot the control action (error in this case) as a function of time, we obtain Figs. 1.4(a) and (b), respectively. In Fig. 1.4(a) it is interesting to see the input as a step input and the plant output as an exponential response. If we let our minds wander, we may recall that this type of response is also exhibited in a low-pass filter with a step excitation. In a frequency sense

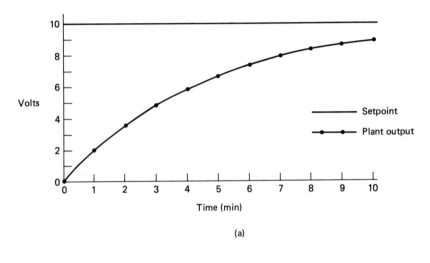

(a)

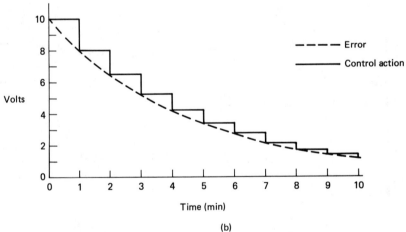

(b)

Figure 1.4   (a) Setpoint, plant output vs. time, (b) error vs. time

we are saying that our plant is allowing a band of frequencies to pass and is attenuating a considerable number of frequencies. This arises from the fact that our step input can be disassembled into its harmonic components and that all waveforms can be shown as a Fourier series. As we mentioned earlier, an increase in gain in the compensation network will promote faster response in our control system. In a frequency sense it appears that our bandwidth is increasing, where the bandwidth represents the frequencies passed by our plant. It would appear that we could design or analyze our control system in a frequency domain which would be independent of time. This domain is the vehicle by which we will travel through our control system in the chapters to come.

In Fig. 1.4(b) we find the control action (or error in this case) to be decreasing with time. You were asked whether this control system is continuous, and from Fig. 1.4(b) it is evident that the control action is discrete, making the control system discrete, and therefore not continuous. If the control action were modified at all times, we would attain continuous control.

It is important to understand that the plant output is always continuous, since physical components are involved. The idea of a discrete control system indicates the discrete nature of the control action only. The control action in Example 1.1 was discretized for ease of calculation and to generate a finite set of data points for ease of illustration. This form of control represents the method of computerized control, and it is safe to say that human beings attack a problem in the same fashion as a computer with rather limited speed.

To attach a name to this kind of control system implies there is at least one other kind. Let's define our present control system and then pursue the other.

1.2 CLOSED-LOOP CONTROL

Example 1.1 was an example of **closed-loop control**. This type of control system requires negative feedback to create an error and subsequent control action. The plant responds with an output which is returned via negative feedback, and the process continues in an endless loop. The following are the main benefits of closed-loop control:

1. Increases accuracy.
2. Reduces effects of noise.

3. Increases plant bandwidth.
4. Can be applied with minimal input/output relationships of the plant under control.

There are two disadvantages in terms of practical implementation of this control system:

1. High cost.
2. Requires qualified personnel to ensure stability.

1.3 OPEN-LOOP CONTROL

Open-loop control requires no feedback and therefore is not error based. To make the control system of Example 1.1 an open-loop system we would simply remove the feedback path and eliminate the transducer. Now to obtain our proper level in the tank a 10-volt signal at the valve for 5 minutes would suffice. Owing to the elimination of the feedback mechanism, this system has several advantages:

1. Less expensive than closed-loop.
2. Easy to design.
3. Minimal stability problems.

The disadvantages are that it

1. Requires an accurate input/output relationship of the plant.
2. Does not measure error, thus assumes that output is correct.

The accuracy of open-loop control depends on the accuracy of the input/output relationship of the plant. If the latter is adequate and the process will not exhibit any odd behavior in its response, then this type of control can be readily applied. A good example of an open-loop control system can be found in line-feed mechanisms of small dot-matrix printers. The movement is initiated via a stepper motor which has a defined input/output relationship and is guaranteed a certain accuracy. This accuracy can be attained without any feedback only if the system responds correctly and without external intervention. What happens if the stepper motor loses a few steps? Is there any way of correcting this? What about feedback? This brings us back to closed-loop control.

We have mentioned open-loop control here only in passing; our focus in this text will be on closed-loop control.

1.4 SUMMARY

We have represented the control system in a very nonmathematical way and have looked at both closed- and open-loop control.   In the former we found that an error was generated from the sum of the setpoint and the inverted feedback (negative feedback) of the plant output.   This error was processed by the compensation network to generate a control action.   This in turn provided the driving force for the plant, which in turn modified the plant output.   This process continued until the error diminished.   The type of control action determined whether the control system was continuous or discrete.   We pondered the idea of using a frequency domain in the design and analysis of closed-loop control systems.

In open-loop control we found that no feedback was present and the accuracy was directly related to the accuracy of the input/output relationship of the plant.   We mentioned open-loop control only in passing; our focus in this text will be on closed-loop control systems.

Problems

1.1. Give examples of continuous control systems.

1.2. "If the feedback in our control system were added to our setpoint, then our system would have positive feedback and tend to instability."   Comment on this statement.

1.3. A control system is operating and the feedback is accidentally disconnected. What is the effect on the system output?

1.4. Repeat the control process as indicated in Example 1.1 with the compensating network multiplying the error by 5.   How long does it take to fill the tank? Has the bandwidth of our system increased?

1.5. The control system in Example 1.1 has been accidentally modified.   The modification consists of a puncture in the tank.   You can assume that the liquid leaving is constant and equivalent to 1 meter per minute in terms of the tank height.   If the compensating network multiplies the error by 10, redo the control process taking into consideration the hole in the tank.   How long does it take to fill the tank? What is the final voltage on the proportional valve?

1.6. The proportional valve in Example 1.1 is faulty. The circuitry of the valve adds a constant 20% of the applied voltage. That is, if the input voltage to the valve is 10 V, then the valve assumes that 12 V is present. If the compensating network multiplies the error by 4, redo the control process. What is the error in the final output of the control system? If an open-loop system were used, what would be the percentage error in the tank level?

1.7. Repeat Problem 1.6, assuming that the transducer is also faulty. The transducer has lost calibration and is reading 20% high.

1.8. Write a computer program (in a language of your choice) that performs the control process as outlined in Example 1.1. It should be user friendly and ask for the system setpoint and compensating gain. The values of Table 1.1 should be generated to verify its operation. As system simulation is very important, the reader can also graphically display the control process.

Part I

Continuous

2

Laplace Transforms

2.1 INTRODUCTION

In Chapter 1 we were introduced to the closed-loop control system in a very nonmathematical manner. As we begin to analyze this type of control system, we need to have the proper mathematical tools at our disposal.

Since all waveforms can be represented as a Fourier series, it is evident that a time-domain study would be cumbersome and labor intensive, since an infinite number of harmonics would have to be considered. If time is eliminated and a frequency base used, then the analysis becomes manageable and more representative.

We will find that the Laplace transform is the tool required to perform the time-to-frequency transformation. We will also find that differential equations are transformed to algebraic equations in the frequency domain. Why is this significant? The differential equation represents the governing equation of all physical systems, and the solution provides the output relative to a given input. This in fact is a control system, as shown in the previous chapter.

In this chapter we will introduce the Laplace transform and generate the required transforms. We will introduce the inverse Laplace transform, utilizing both partial fractions and the convolution integral as tools for the transfer. We will generate and solve differential equations using Laplace transforms. To facilitate the simulation of systems in real time we will introduce the operational amplifier as a tool for simulating physical systems. This analog simulation will be invaluable in comparing the predicted and actual output of our control system.

2.2 DERIVATION OF REQUIRED LAPLACE TRANSFORMS

Early in the nineteenth century a French mathematician and astronomer, Pierre Simon de Laplace, developed the **Laplace transform** as a means for the solution of differential equations. At that time it was strictly a mathematical exercise, and not until the twentieth century was it applied to the field of electronic and controls technology. Equation (2.2.1) represents the Laplace transform of a time function $f(t)$:

$$F(s) = \mathcal{L}[f(t)] \triangleq \int_0^\infty f(t) \cdot e^{-st}\, dt \qquad (2.2.1)$$

It is apparent that this transform multiplies our time function by an exponentially decaying function. The time integration uniquely transforms our time function into a new function of the variable s. If we let $s = j\omega$ ($\omega = 2\pi f$), then our new function is in a frequency domain. (By convention, all time functions are in lower case while Laplace functions are in capitals. The variable s, Laplace operator, is lower case only.) Equation (2.2.1) represents a one-sided Laplace transform, as it assumes that all times are positive—A good assumption for practical systems.

> **EXAMPLE 2.1**   Transform of the Dirac delta ($\delta(t)$) function
>
> The **Dirac delta function** is a mathematical wonder as it has a width of zero, a height of infinity, and an area of unity. It is important in the discrete control system as it represents the idea of sampling. If you have difficulty visualizing this function, consider Fig. 2.1. We see that as the width of the rectangle decreases to zero, the height increases to infinity in order to maintain an area of unity. Using equation (2.2.1), we obtain the Laplace transform as follows:
>
> $$\mathcal{L}[\delta(t)] = \int_0^\infty \delta(t) \cdot e^{-st}\, dt \qquad (2.2.2)$$
>
> If we consider that the Dirac delta exists at $t = 0$ only, equation (2.2.2)

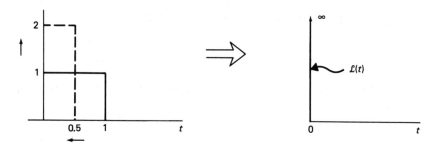

Figure 2.1   Dirac delta function

becomes

$$\mathcal{L}[\delta(t)] = \int_0^\infty \delta(t) \cdot 1 \, dt \qquad (2.2.3)$$

which represents the area under the curve and by definition is equal to unity. Therefore the Laplace transform of the Dirac delta function (sometimes known as **unit impulse**) is independent of the Laplace operator s. The result is

$$\mathcal{L}[\delta(t)] = 1 \qquad (2.2.4)$$

EXAMPLE 2.2   Transform of the step ($u(t)$) function

In Chapter 1 we found that the input or setpoint of a control system was in the form of a step function.   Equation (2.2.5) provides the mathematical definition and Fig. 2.2 the graphical representation of the step function ($u(t)$).

$$\begin{cases} u(t) = 0, & t < 0 \\ u(t) = 1, & t \ge 0 \end{cases} \qquad (2.2.5)$$

Using equation (2.2.1), we obtain the transform of the step function.

$$\mathcal{L}[u(t)] = \int_0^\infty 1 \cdot e^{-st} \, dt \qquad (2.2.6)$$

Figure 2.2   Step function

and, performing the integration, we obtain

$$\mathcal{L}[u(t)] = \left| -\frac{1}{s} e^{-st} \right|_0^\infty \tag{2.2.7}$$

By applying limits of integration we obtain the required transform:

$$\mathcal{L}[u(t)] = \frac{1}{s} \tag{2.2.8}$$

EXAMPLE 2.3 Transform of an exponentially decaying function

Equation (2.2.9) represents the time function $f(t)$.

$$f(t) = e^{-at} \tag{2.2.9}$$

Using equation (2.2.1), we obtain the transform of the time function:

$$\mathcal{L}[f(t)] = \int_0^\infty e^{-at} \cdot e^{-st} \, dt \tag{2.2.10}$$

and, performing the integration, we obtain

$$\mathcal{L}[f(t)] = \left| -\frac{1}{s+a} e^{-(s+a)t} \right|_0^\infty \tag{2.2.11}$$

By applying the limits of integration we generate the required transform:

$$\mathcal{L}[e^{-at}] = \frac{1}{s+a} \tag{2.2.12}$$

EXAMPLE 2.4 Transform of sin (ωt) and cos (ωt)

At first glance this transform would require integration by parts and for the mathematical enthusiast would provide ample challenge. It is evident that exponential functions facilitate easy integration. If we use Euler's identity (2.2.13), we find that our sinusoidal waveforms can in fact be represented in exponential form:

$$f(t) = e^{j\omega t} = \cos(\omega t) + j \sin(\omega t) \tag{2.2.13}$$

If we transform equation (2.2.13), it is evident that we will have generated the transforms of both sin (ωt) and cos (ωt). Since the Laplace transform is a linear transformation, we can say that

$$\mathcal{L}[\sin(\omega t)] = \text{Im} \{\mathcal{L}[f(t)]\} \tag{2.2.14}$$

and

$$\mathcal{L}[\cos(\omega t)] = \text{Re}\{\mathcal{L}[f(t)]\} \qquad (2.2.15)$$

Using equation (2.2.1), we obtain the transform for $f(t)$:

$$\mathcal{L}[f(t)] = \int_0^\infty e^{j\omega t} \cdot e^{-st} \, dt \qquad (2.2.16)$$

and, performing the integration, we obtain

$$\mathcal{L}[f(t)] = \left| -\frac{1}{s - j\omega} e^{-(s - j\omega)t} \right|_0^\infty \qquad (2.2.17)$$

Applying the limits of integration, multiplying by the complex conjugate, and simplifying, we obtain the required transform:

$$\mathcal{L}[f(t)] = \frac{s}{s^2 + \omega^2} + j\frac{\omega}{s^2 + \omega^2} \qquad (2.2.18)$$

Applying Equations (2.2.14) and (2.2.15), we obtain the transforms of sin (ωt), cos (ωt) by extracting the imaginary and real parts, respectively, of equation (2.2.18):

$$\mathcal{L}[\sin(\omega t)] = \frac{\omega}{s^2 + \omega^2} \qquad (2.2.19)$$

$$\mathcal{L}[\cos(\omega t)] = \frac{s}{s^2 + \omega^2} \qquad (2.2.20)$$

EXAMPLE 2.5   Transform of the exponentially damped sine and cosine functions

As in the previous example, we represent these functions in exponential form.   Equation (2.2.21) provides this form:

$$f(t) = e^{-at} \cdot e^{j\omega t} = e^{-at}(\cos(\omega t) + j\sin(\omega t)) \qquad (2.2.21)$$

It is left as an exercise for the reader to generate the results given below.

$$\mathcal{L}[e^{-at}\sin(\omega t)] = \frac{\omega}{(s + a)^2 + \omega^2} \qquad (2.2.22)$$

$$\mathcal{L}[e^{-at}\cos(\omega t)] = \frac{s + a}{(s + a)^2 + \omega^2} \qquad (2.2.23)$$

The foregoing examples have covered the most important Laplace transforms encountered in control systems.   To make them accessible for future use they are summarized in Table 2.1.

TABLE 2.1   Table of Laplace Transforms

	Time function	Laplace transform
Dirac delta	$\delta(t)$	1
Step	$u(t)$	$\dfrac{1}{s}$
Damped exponential	e^{-at}	$\dfrac{1}{s + a}$
Sine	$\sin(\omega t)$	$\dfrac{\omega}{s^2 + \omega^2}$
Cosine	$\cos(\omega t)$	$\dfrac{s}{s^2 + \omega^2}$
Damped sine	$e^{-at}\sin(\omega t)$	$\dfrac{\omega}{(s + a)^2 + \omega^2}$
Damped cosine	$e^{-at}\cos(\omega t)$	$\dfrac{s + a}{(s + a)^2 + \omega^2}$

2.3 PROPERTIES OF LAPLACE TRANSFORMS

Linearity Property

As stated earlier, the Laplace transform is a linear transformation. This **linearity** property implies that a sum of time functions can be transformed as the sum of the individual transforms. Mathematically this is equivalent to

$$\mathcal{L}[a_1 f_1(t) + a_2 f_2(t)] \tag{2.3.1}$$

which simplifies to

$$a_1 \mathcal{L}[f_1(t)] + a_2 \mathcal{L}[f_2(t)] \tag{2.3.2}$$

EXAMPLE 2.6

Given the following time function, use the table of transforms to generate the required transform:

$$f(t) = 2 + 3e^{-t} - 3e^{-2t} \cos 6t$$

> Using the linearity property, we can say that
>
> $$\mathcal{L}[f(t)] = \mathcal{L}[2] + 3\mathcal{L}[e^{-t}] - 3\mathcal{L}[e^{-2t}\cos 6t]$$
>
> and, using Table 2.1, generate the required transform:
>
> $$\mathcal{L}[f(t)] = \frac{2}{s} + \frac{3}{s+1} - \frac{3(s+2)}{(s+2)^2 + 36}$$

A common misunderstanding occurs when two time functions are multiplied together and it is assumed that the transform is the multiplication of the individual transforms. This assumption is incorrect, as a brief reflection on equation (2.2.1) suggests that a complex convolution integral exists.

Derivative Property

The **derivative** property is very important for the transformation of differential equations into the Laplace domain. For ease of calculation we will consider all initial conditions to be zero. As a result, our form of this property will not conform to the norm as found in mathematical handbooks. This property states that the nth derivative of a time function appears as the Laplace transform of the time function multiplied by the nth power of the Laplace operator s. Mathematically this is equivalent to

$$\mathcal{L}\left[\frac{d^n}{dt^n} f(t)\right] \tag{2.3.3}$$

which simplifies to

$$s^n \cdot F(s) \tag{2.3.4}$$

> **EXAMPLE 2.7**
>
> Find the Laplace transform of the following time function, assuming all initial conditions are equal to zero:
>
> $$\mathcal{L}\left[\frac{d^3}{dt^3} f(t) + \frac{d}{dt} f(t)\right]$$
>
> Using the derivative property, we obtain the required result:
>
> $$s^3 \cdot F(s) + s \cdot F(s)$$

Integral Property

The **integral** property allows the transformation of a time function that is being integrated. To transform an integral function, we simply divide the Laplace transform of the time function being integrated by the Laplace operator s. Mathematically this is equivalent to

$$\mathscr{L}\left[\int_0^t f(\tau)\ d\tau\right] \tag{2.3.5}$$

which simplifies to

$$\frac{F(s)}{s} \tag{2.3.6}$$

Final-Value Theorem

The **final-value theorem** is very important to control systems. It allows us to obtain the steady-state value of our output while in the Laplace domain. Mathematically this translates to the limit of our time function as time approaches infinity. In the Laplace domain this is equivalent to the limit of the Laplace transform multiplied by the Laplace operator s as the Laplace operator approaches zero. That is,

$$\lim_{t\to\infty} f(t) = \lim_{s\to 0} s \cdot F(s) \tag{2.3.7}$$

EXAMPLE 2.8

The steady-state value of a time function is required and the Laplace transform of the time function is available. The transform is

$$F(s) = \frac{4}{s(s+2)}$$

Using the final-value theorem, we obtain the required result:

$$\lim_{t\to\infty} f(t) = \lim_{s\to 0} s \cdot \frac{4}{s(s+2)} = 2$$

Shifting Property

The **shifting** property is also important in that it allows the incorporation of *transport lag* or *dead time* into the Laplace domain. This phenomenon occurs in control systems and is basically the time delay between input change and plant response.   If a time function is delayed a time units, we have

$$g(t) = \begin{cases} f(t - a), & t > a \\ 0, & t < a \end{cases} \qquad (2.3.8)$$

Using the shifting property we generate the Laplace transform

$$\mathscr{L}[g(t)] = e^{-as} \cdot G(s) \qquad (2.3.9)$$

EXAMPLE 2.9

A control action of 4 units is applied to a control system, and Fig. 2.3 represents the time lag.   Find the Laplace transform of the control action.

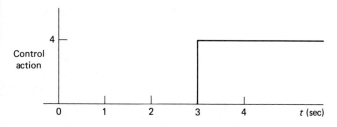

Figure 2.3 Control action for control system

From Fig. 2.3 we find the control action as a step function of 4 units delayed 3 seconds (dead time).   Using the shifting property, we obtain the required Laplace transform:

$$\frac{4e^{-3s}}{s}$$

This concludes the introduction of properties of Laplace transforms as they apply to control systems.   You should be sure you fully understand these properties before proceeding, as they will be used throughout the text.

2.4 INVERSE LAPLACE TRANSFORM

In the preceding section we transformed our time functions into the Laplace domain.   While this will simplify our analysis of control systems, we should still be able to see the time response.   This implies a transformation from the Laplace to the time domain, or the **inverse Laplace transform**.   The two most common techniques used are the *partial-fraction expansion* and the *convolution integral*.   The first makes use of algebraic manipulation followed by a revisit to Table 2.1, while the second revisits Table 2.1 and then performs an integration process.   (Since both techniques use Table 2.1, it is important that our Laplace transforms appear as in this table for easy access to the time function.)

Partial-Fraction Technique

It is important to note that the **partial-fraction technique** will operate only on Laplace transforms with a numerator that is of lower order than the denominator.   In control systems we are concerned with two forms of transforms and will cover these in two separate cases.

Case 1.   Consider the Laplace transform $F(s)$ as given below:

$$F(s) = \frac{a_{n-1}s^{n-1} + a_{n-2}s^{n-2} + \cdots + a_1 s + a_0}{b_n s^n + b_{n-1}s^{n-1} + \cdots + b_1 s + b_0} \qquad (2.4.1)$$

where $a_0, \cdots, a_{n-1}, b_0, \cdots, b_n, n$ are constants.   If the denominator contains real roots only, we can rewrite equation (2.4.1) as follows:

$$F(s) = \frac{a_{n-1}s^{n-1} + a_{n-2}s^{n-2} + \cdots + a_1 s + a_0}{(s + p_n)(s + p_{n-1}) \cdots (s + p_0)} \qquad (2.4.2)$$

where $p_n, \ldots, p_0$ are appropriate constants.   Mathematically we can disassemble equation (2.4.2) as follows:

$$F(s) = \frac{A_n}{(s + p_n)} + \frac{A_{n-1}}{(s + p_{n-1})} + \cdots + \frac{A_0}{(s + p_0)} \qquad (2.4.3)$$

where $A_n, \ldots, A_0$ are constants to be calculated.   If we take a common denominator in equation (2.4.3) we obtain

$$F(s) = \frac{B_{n-1}s^{n-1} + B_{n-2}s^{n-2} + \cdots + B_1 s + B_0}{(s + p_n)(s + p_{n-1}) \cdots (s + p_0)} \qquad (2.4.4)$$

where $B_{n-1}, \ldots, B_0$ are functions of $A_n, \ldots, A_0$.   Now equations (2.4.4) and (2.4.2) represent the same expression, implying the following set of equations:

$$\begin{cases} B_{n-1} = a_{n-1} \\ B_{n-2} = a_{n-2} \\ \quad\vdots \\ B_0 \;\;\;= a_0 \end{cases} \qquad (2.4.5)$$

The solution of the n simultaneous equations in (2.4.5) will yield the coefficients required in equation (2.4.3). A close inspection of equation (2.4.3) and Table 2.1 indicates that the following time function exists:

$$\mathcal{L}^{-1}[F(s)] = f(t) = A_n e^{-p_n t} + A_{n-1} e^{-p_{n-1}t} + \cdots + A_0 e^{-p_0 t} \qquad (2.4.6)$$

The inverse Laplace transform also enjoys the property of linearity, as indicated by the previous operation. An example is past due.

EXAMPLE 2.10

Given the following Laplace transform $F(s)$, find the corresponding time function by applying the inverse Laplace transform $\{\mathcal{L}^{-1}[F(s)]\}$:

$$F(s) = \frac{s + 4}{s(s^2 + 3s + 2)}$$

First we rewrite our transform as given in equation (2.4.2). It is evident that the roots in the denominator are in fact real.

$$F(s) = \frac{s + 4}{s(s + 2)(s + 1)} \qquad (1)$$

We now disassemble equation (1) as shown in equation (2.4.3):

$$F(s) = \frac{A_2}{s} + \frac{A_1}{s + 2} + \frac{A_0}{s + 1} \qquad (2)$$

Taking a common denominator in equation (2), we obtain

$$F(s) = \frac{s^2(A_2 + A_1 + A_0) + s(3A_2 + A_1 + 2A_0) + 2A_2}{s(s + 2)(s + 1)} \qquad (3)$$

and, comparing equations (3) and (1), we have the following simultaneous equations:

$$A_2 + A_1 + A_0 = 0 \qquad (4)$$

$$3A_2 + A_1 + 2A_0 = 1 \qquad (5)$$

$$2A_2 + 0A_1 + 0A_0 = 4 \qquad (6)$$

Solving the simultaneous equations and substituting into equation (2), we obtain

$$F(s) = \frac{2}{s} + \frac{1}{s + 2} - \frac{3}{s + 1} \qquad (7)$$

From equation (7) and Table 2.1 we obtain the required time function ($f(t)$):

$$f(t) = \mathcal{L}^{-1}[F(s)] = 2 + e^{-2t} - 3e^{-t}$$

Case 2. This case is very similar to case 1, differing only in that there is at least one imaginary root in the denominator of the transform. An example will indicate the deviation in procedure.

EXAMPLE 2.11

Given the following Laplace transform $F(s)$, find the corresponding time function by applying the inverse Laplace transform $\{\mathcal{L}^{-1}[F(s)]\}$:

$$F(s) = \frac{s + 4}{s(s^2 + 2s + 2)}$$

Upon inspection of the denominator we find the discriminant of the quadratic equation to be less than zero. This implies imaginary roots. At this point we complete the square and leave the quadratic as a single factor:

$$F(s) = \frac{s + 4}{s[(s + 1)^2 + 1]} \tag{1}$$

A quick glance at Table 2.1 and equation (1) will indicate the relevance of completing the square. It appears that our time function will have terms containing damped sine and cosine terms. We now disassemble equation (1) as shown in equation (2.4.3) with one exception. The term containing the quadratic denominator requires a numerator of one order less in terms of the Laplace operator (s). Let's continue:

$$F(s) = \frac{A_2}{s} + \frac{A_1 s + A_0}{(s + 1)^2 + 1} \tag{2}$$

Taking a common denominator in equation (2), we obtain

$$F(s) = \frac{s^2(A_2 + A_1) + s(2A_2 + A_0) + 2A_2}{s[(s + 1)^2 + 1]} \tag{3}$$

and, comparing equations (3) and (1), we have the following simultaneous equations:

$$A_2 + A_1 + 0A_0 = 0 \tag{4}$$

$$2A_2 + 0A_1 + A_0 = 1 \tag{5}$$

$$2A_2 + 0A_1 + 0A_0 = 4 \tag{6}$$

Solving the simultaneous equations and substituting into equation (2), we obtain

$$F(s) = \frac{2}{s} - \frac{2s + 3}{(s + 1)^2 + 1} \tag{7}$$

In order that Table 2.1 can readily be used, we must first put equation (7) in proper form.   Upon rearrangement, we have

$$F(s) = \frac{2}{s} - 2\left[\frac{s + 1}{(s + 1)^2 + 1} + 0.5\frac{1}{(s + 1)^2 + 1}\right] \qquad (8)$$

The form given in equation (8) is imperative in order that Table 2.1 can be utilized.   The reader is asked to make sure that this form is clear and to review the process until it is so.   From Table 2.1 and equation (8) we obtain the required time function ($f(t)$):

$$f(t) = \mathscr{L}^{-1}[F(s)] = 2 - 2e^{-t}\cos t - e^{-t}\sin t$$

In subsequent chapters it will be necessary to deal with cubic expressions in the denominator of the Laplace transforms.   If we consider a cubic equation, we find that there is at least one real root, and if it is found, the remaining quadratic equation can be obtained using long division.   At this point either case 1 or case 2 will apply, and we can perform the inverse Laplace transform. Figure 2.4 displays a sample BASIC program that extracts at least one real

```
10  PRINT "THIS PROGRAM WILL SEPARATE A THIRD ORDER"
20  PRINT "FUNCTION INTO A FIRST AND SECOND ORDER"
30  PRINT "FUNCTION. THE EQUATION WILL BE IN THE"
40  PRINT "FORM : s^3 + A*s^2 + B*s + C"
50  INPUT "A = ",A
60  INPUT "B = ",B
70  INPUT "C = ",C
80  UPPER = 100
90  LOWER = -100
100 TRIAL = (UPPER + LOWER)/2
110 VALUE = TRIAL^3 + A*TRIAL^2 + B*TRIAL + C
120 IF VALUE>0 THEN UPPER = TRIAL ELSE LOWER = TRIAL
130 IF VALUE<.0001 AND VALUE>-.0001 THEN 160
140 IF UPPER = LOWER THEN 160
150 GOTO 100
160 J = TRIAL + A
170 K = B + TRIAL*(TRIAL + A)
180 IF J>0 THEN B$ = "+" ELSE B$ = "-"
190 IF K>0 THEN C$ = "+" ELSE B$ = "-"
200 IF TRIAL<0 THEN A$ = "+" ELSE A$ = "-"
210 PRINT
220 PRINT "FIRST ORDER SYSTEM"
230 PRINT "s ";A$;" ";ABS(TRIAL)
240 PRINT
250 PRINT "SECOND ORDER SYSTEM"
260 PRINT "s^2 ";B$;" ";ABS(J);"*s ";C$;" ";ABS(K)
270 END
```

Figure 2.4   BASIC program to disassemble a cubic equation

root from a cubic equation and performs the long division, generating the remaining quadratic equation. The program utilizes a binary search technique and minimizes the mathematical hardship.

EXAMPLE 2.12

Given the following Laplace transform $F(s)$, find the corresponding time function by applying the inverse Laplace transform $\{\mathcal{L}^{-1}[F(s)]\}$:

$$F(s) = \frac{s + 3}{s(s^3 + 10s^2 + 27s + 24)} \qquad (1)$$

Using the provided BASIC program, we obtain

$$F(s) = \frac{s + 3}{s(s + 6.337)(s^2 + 3.663s + 3.787)} \qquad (2)$$

From equation (2) it is evident that the roots of the quadratic equation are imaginary and that case 2 will apply. It is left as an exercise for the reader to show that the time function is

$$\mathcal{L}^{-1}[F(s)] = f(t) = 0.125 + 0.0254e^{-6.337t}$$

$$- 0.1504e^{-1.832t}(\cos 0.658t + 1.158 \sin 0.658t)$$

This concludes the partial-fraction method.

Convolution-Integral Technique

The **convolution-integral technique** is popular in modern control theory and state-variable analysis; it is included here to expose the reader to that area. The convolution integral is

$$\mathcal{L}^{-1}[F_1(s) \cdot F_2(s)] = \int_0^t f_1(\tau) \cdot f_2(t - \tau) \, d\tau \qquad (2.4.7)$$

An example will help the reader appreciate this process.

EXAMPLE 2.13

Apply the convolution integral to the following Laplace transform in order to generate the inverse Laplace transform or time function $f(t)$:

$$F(s) = \frac{s}{(s + 1)(s + 2)} \qquad (1)$$

SOLUTION   Now equation (1) can be shown as

$$F(s) = s\left(\frac{1}{s+1}\right)\left(\frac{1}{s+2}\right) \tag{2}$$

The first factor implies a time derivative, as we recall the properties of the Laplace transform. The two remaining factors become the transforms $F_1(s)$ and $F_2(s)$, and a quick look at Table 2.1 reveals the time functions $f_1(t)$ and $f_2(t)$, which are

$$f_1(t) = e^{-t} \tag{3}$$

$$f_2(t) = e^{-2t} \tag{4}$$

Using equation (2.3.7) and noting that we require a time derivative, we generate the inverse Laplace transform

$$\mathcal{L}^{-1}[F(s)] = f(t) = \frac{d}{dt}\left(\int_0^t e^{-\tau} \cdot e^{-2(t-\tau)}\, d\tau\right)$$

which, upon integration and subsequent differentiation, simplifies to

$$f(t) = 2e^{-2t} - e^{-t}$$

The reader is advised to use partial fractions on the transfer function given by expression (1) as a means of verifying the results obtained by convolution.

If we rewrite equation (2.4.2) as factors of transforms, we obtain

$$F(s) = (a_{n-1}s^{n-1} + a_{n-2}s^{n-2} + \cdots + a_0)$$

$$[F_n(s)][F_{n-1}(s)] \cdots [F_0(s)] \tag{2.4.8}$$

and, applying the derivative property and convolution, we obtain the time function ($f(t)$):

$$f(t) = \left(a_{n-1}\frac{d^{n-1}}{dt^{n-1}} + a_{n-2}\frac{d^{n-2}}{dt^{n-2}} + \cdots + a_0\right)$$

$$(f_n * f_{n-1} * \cdots * f_1 * f_0) \tag{2.4.9}$$

where the asterisk indicates the convolution operator. In the general case the time function requires n convolutions with $n - 1$ differentiations and algebraic sums. The reason for the n convolutions is apparent in expression (2.4.7), as only two time functions convolve at one time. The order of convolution is of no importance, as it is commutative. If any integration involves sine or cosine terms, it is generally handled with Euler's identity. This provides a homogeneity in the integral process, as all functions are exponential in nature. This technique was used in the derivation of the Laplace transforms for the sine and cosine functions earlier in this chapter.

We now have acquired all the mathematical tools used to transform our control system into the Laplace domain. At this point we will investigate the solution of some physical systems by generating the governing differential equations and solving them using Laplace transforms. This will provide the reader with some insight as to the importance of Laplace transforms and their place in the simplification process.

2.5 DIFFERENTIAL EQUATIONS

Most physical systems can be explained in terms of differential equations with constant coefficients. Let us investigate some of those systems and see how Laplace transforms can expedite the solution process. As before, we will explore with examples.

EXAMPLE 2.14

Figure 2.5(a) provides a simple *RC* electrical circuit with a constant voltage supply v, current $i(t)$, and switch S1, which is open. Figure 2.5(b) displays the same circuit as an impedance network. At time $t = 0$ the switch is closed and we are to find the current flow $i(t)$ for all time.

SOLUTION   Using Kirchhoff's voltage law, we have

$$v = R \cdot i(t) + \frac{1}{C} \int i(t)\, dt \qquad (1)$$

If we apply the properties of Laplace transforms to differential equation (1), we obtain

$$\frac{V}{s} = R \cdot I(s) + \frac{1}{sC} I(s) \qquad (2)$$

From a glance at equation (2) it is evident that the new Laplace equation is

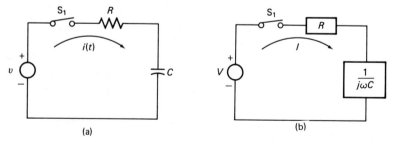

(a)   (b)

Figure 2.5   *RC* circuit (a) in time, (b) as an impedance network

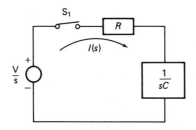

Figure 2.6 Laplace equivalent of *RC* circuit

algebraic. If we use equation (2) to generate the new Laplace circuit, we obtain Fig. 2.6.

Comparing Fig. 2.6 with Fig. 2.5(b), we find

$$s = j\omega \tag{2.5.1}$$

From equation (2.5.1) it is evident that our circuit is in the frequency domain. Now back to the problem. Solving for $I(s)$ in equation (2), we have

$$I(s) = \frac{\dfrac{V}{R}}{s + \dfrac{1}{RC}} \tag{3}$$

To find the solution in time ($i(t)$) we require the inverse Laplace transform of equation (3). From Table 2.1 we have

$$i(t) = \frac{v}{R} \cdot e^{-t/RC} \tag{4}$$

which represents the time solution for the current in the circuit. In essence, we have solved a differential equation with simple algebraic manipulation. The reader is asked to repeat this problem with an inductor (L) as opposed to a capacitance. If $1/sC$ is the equivalent of a capacitor in the Laplace domain, what is the equivalent of the inductor? From equation (2) we find that our supply voltage v was shown as V/s in the Laplace domain. Why? Consider the switch closing at time $t = 0$ and think about the entries in Table 2.1.

EXAMPLE 2.15

Figure 2.7 portrays the physical makeup of many mechanical mechanisms. This system consists of a mass M attached to a fixed spring (spring constant k). An external force $u(t)$ moves the mechanism a distance x. If the force applied to the mechanism is a unit impulse (Dirac delta), trace the path of movement.

SOLUTION   Using the laws of physics, we find that the force exhibited by

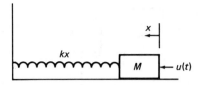

Figure 2.7   Spring-mass mechanism

the mass M is equivalent to the mass times its acceleration. The spring applies a force equivalent to its spring constant times the distance travelled. The sum of these forces must equal the applied force, or mathematically

$$M\frac{d^2x}{dt^2} + kx = \delta(t) \tag{1}$$

If we apply the properties of Laplace transforms to differential equation (1), we obtain

$$Ms^2X(s) + kX(s) = 1 \tag{2}$$

and, solving for $X(s)$, we have

$$X(s) = \frac{1}{Ms^2 + k} \tag{3}$$

If we put equation (3) in proper form, we have

$$X(s) = \frac{1}{\sqrt{kM}}\frac{\sqrt{\dfrac{k}{M}}}{s^2 + \dfrac{k}{M}} \tag{4}$$

which allows us to use Table 2.1 to find the inverse Laplace transform:

$$x(t) = \frac{1}{\sqrt{kM}}\sin\left(\sqrt{\frac{k}{M}}\right)t \tag{5}$$

which happens to be the path of movement in time. If we look at equation (5), it is apparent that the movement is sinusoidal and implies perpetual motion. How convenient! Actually this is a deliberate mistake that has two purposes: (1) awakening the reader, and (2) providing another intellectual exercise for the reader.

It is obvious that we forgot to consider the friction in the system and it should be apparent that our solution, though simple, is incorrect. One should guarantee that the governing differential equation is correct and at the same time have some idea of system response. In these times of high-speed computational machines and higher-order mathematics we tend to bypass understanding. These are mere tools, however, that enhance but do not replace the same.

Now that you have been intellectually awakened, you should redo the problem with consideration of friction in the system. The friction of the mass is proportional to the speed of the mass times a constant B and will appear on the left-hand side of equation (1). Is the motion still perpetual?

EXAMPLE 2.16

A physical system is found to have the following differential equation with all initial conditions being zero:

$$\ddot{x} + 4\dot{x} + 3x = u(t); \qquad \dot{x} = \frac{d}{dt} \qquad (1)$$

If the input ($u(t)$) is given by equation (2), find the system response ($x(t)$).

$$u(t) = 2 \cos 3t \qquad (2)$$

SOLUTION   If we take Laplace transforms of both equations and combine them, we obtain

$$X(s) = \frac{2s}{(s^2 + 9)(s^2 + 4s + 3)} \qquad (3)$$

Applying partial fractions and solving for the coefficients, we obtain the usable form

$$X(s) = \frac{-0.09s}{s^2 + 9} + 0.121 \frac{3}{s^2 + 9} + \frac{0.136}{s + 3} - \frac{0.046}{s + 1} \qquad (4)$$

To obtain the time response $x(t)$ we simply find the inverse Laplace transform of expression (4) using Table 2.1:

$$x(t) = 0.09 \cos 3t + 0.121 \sin 3t + 0.136e^{-3t} - 0.046e^{-t} \qquad (5)$$

It will benefit the reader to plot the input $u(t)$ and system output $x(t)$ with respect to time. From the plot the reader should consider whether the system is in fact responding to the input or whether the input is beyond the frequency response of the system. An awareness of the answer will illuminate the importance of the bandwidth of a physical plant or component.

It is evident that Laplace transforms simplify the solutions of differential equations. Since control systems are governed by differential equations, Laplace transforms will simplify the analysis of the same. In the design process it is convenient to simulate our plant or entire control system. This nondestructive testing is both inexpensive and fast. We now turn our attention to this simulation process.

2.6 ANALOG SIMULATION

Analog simulation, as the name suggests, involves electronic circuitry. The basic building block for analog simulation is provided by the operational amplifier. Figure 2.8 is a circuit diagram for the typical configuration of the operational amplifier that will be used. We will assume ideal operational amplifiers with negative feedback, which enjoy the following characteristics:

1. Voltages at both inputs are equal.
2. Inputs require no current.

Using these characteristics, we can obtain the output of the circuit as given in Fig. 2.8. From characteristic 1 the voltages at both inputs are zero as the positive input is grounded. From characteristic 2 we find that all the current that flows through impedance Z_1 also flows through impedance Z_2. Figure 2.9 reflects this information.

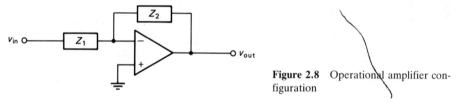

Figure 2.8 Operational amplifier configuration

From Fig. 2.9 we find the output to be equivalent to the voltage drop across impedance Z_2 with reference to the voltage at the negative terminal of the operational amplifier. Since this reference is at zero volts, we find the output at a negative voltage. Mathematically this is given by

$$v_{\text{out}} = -\frac{Z_2}{Z_1} v_{\text{in}} \qquad (2.6.1)$$

It is evident that our output has been inverted and a gain factor exists. Let us investigate two configurations as depicted in Fig. 2.10(a) and (b). The former represents a gain amplifier, the latter an integrating amplifier, both of which are inverting.

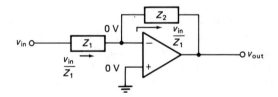

Figure 2.9 Analysis of operational amplifier circuit

Figure 2.10 (a) Grain amplifier, (b) integrating amplifier

If we apply our circuit analysis as before, we obtain the output voltages for the gain and integrating amplifiers, respectively:

$$v_{out} = -\frac{R_2}{R_1} v_{in} \tag{2.6.2}$$

$$v_{out} = -\frac{1}{R_1 C_2} \int v_{in} \, dt \tag{2.6.3}$$

From equation (2.6.3) it is important to note the gain as being

$$\frac{1}{R_1 C_2} \tag{2.6.4}$$

and, if we let $R_1 = 1$ MΩ and $C_2 = 1$ μF, equation (2.6.3) becomes

$$v_{out} = -\int v_{in} \, dt \tag{2.6.5}$$

If we differentiate both sides of equation (2.5.5), we have

$$v_{in} = -\frac{d}{dt} v_{out} \tag{2.6.6}$$

which implies the integrating block as the building block for the solution of a differential equation.   Let's examine an example.

EXAMPLE 2.17

A physical system is found to have the following differential equation with all initial conditions being zero.   It is evident that the system is too expensive to implement and a form of simulation is required for the design process. The design engineer specifies analog simulation.

$$\ddot{x} + 2\dot{x} + 5x = 4 \cdot u(t) \qquad \left(\ddot{x} = \frac{d^2}{dt^2} \right) \tag{1}$$

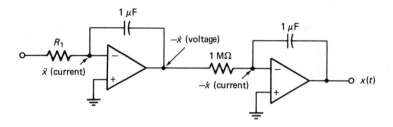

Figure 2.11   First step in simulation

SOLUTION   We rewrite equation (2) so as to isolate the highest differential term—that is,

$$\ddot{x} = 4 \cdot u(t) - 2\dot{x} - 5x \qquad (2)$$

The order of differentiation implies the number of integrating blocks in cascade. We indicate the output of the circuit as the system output $x(t)$ and continue backward with this variable, making sure that expression (2.6.6) is satisfied. It is important to understand that this variable is always available as a current at the inverting input of the operational amplifier and this point is considered to be a summing junction. This process is shown in Fig. 2.11.

The next step is to include the system input $u(t)$. This is connected to R_1 of the first block, and, using equation (2.6.4) with $C_2 = 1\ \mu F$, we calculate the value of R_1 to reflect the gain as indicated in equation (2). Figure 2.12 indicates the second step in the simulation process.

To satisfy equation (2) we require two more terms. Examining Fig. 2.11, we find these terms available but lacking proper gains. To complete the circuit we feed these variables back as voltages, connect them to the current summing junction (indicated by $\ddot{x}$) through resistances calculated by equation (2.6.4) such that equation (2) is satisfied. We find that our output $x(t)$ requires inversion before being fed back, and this is accomplished via the gain amplifier with unity gain. Figure 2.13 provides the completed simulation circuit.

Now that a circuit has been generated, it would be nice to build it and verify its operation. The resistors are $\frac{1}{4}$ watt with a 5% tolerance. The

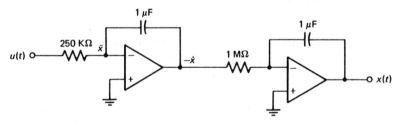

Figure 2.12   Second step in simulation

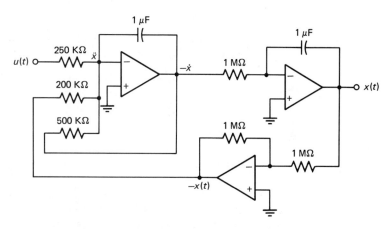

Figure 2.13   Simulation circuit for Example 2.17

capacitors are to be bipolar and preferably of metallized polyester film. The most useful operational amplifier package available, in terms of cost and number of amplifiers, is the LM324N quad amplifier package. Figure 2.14 shows the pin configuration for this device.

From Fig. 2.14 it is evident that the package requires connection of a bipolar power supply, usually ± 15 volts, before the amplifiers can be used. The reader is advised to breadboard the simulation circuit as given in Fig. 2.13. Figure 2.15 shows the overall connections and equipment required.

To ensure initial conditions of zero it is imperative that all capacitors are discharged before switch S_1 is closed. As switch S_1 is closed (step input of 4 volts), the chart recorder or analog voltmeter will display the system step response. It would be nice to know what it is that we are looking for, and this is a good opportunity to apply the Laplace transform.

If we apply Laplace transforms to differential equation (1) of Example 2.17, with zero initial conditions and a step input of 4 units, we obtain

$$X(s) = \frac{4}{s(s^2 + 2s + 5)} \tag{3}$$

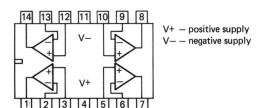

V+ — positive supply
V− — negative supply

Figure 2.14   LM324N quad operational amplifier

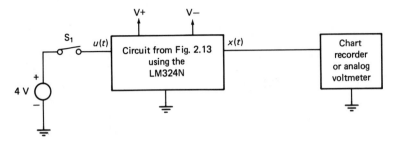

Figure 2.15 Practical circuit simulation diagram

From expression (3) we find that the inverse Laplace transform will have damped sine and cosine terms without any computation. This implies some oscillation before a steady-state value is attained. The steady-state value can be obtained from equation (3) by applying the final-value theorem. This results in a steady-state value of 0.8 units:

$$\lim_{t \to \infty} x(t) = \lim_{s \to 0} s \cdot X(s) = 0.8$$

In terms of your simulation circuit this should result in a steady-state value of 0.8 volt. By quick applications of Laplace transforms we can predict the output and steady-state values. Now we have something to compare to. The reader is advised to complete the Laplace analysis and graph the predicted output. Once the simulation circuit is operational, a valid comparison can be made. How were the results?

   The preceding simulation process is valid for any differential equation with constant coefficients and provides a systematic approach to obtaining the required simulation circuit. To simulate a control system we require one more basic building block, that being the summation block with one input inverted. Figure 2.16(a) shows the block representation, Fig. 2.16(b) the corresponding simulation circuit. By using our two characteristics of

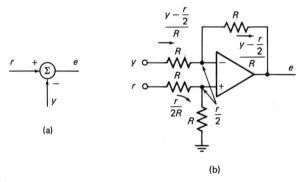

Figure 2.16 Summation: (a) block representation, (b) op-amp circuit

operational amplifiers, we have the output of the circuit as

$$e = \frac{r}{2} - \left(\frac{y - \dfrac{r}{2}}{R}\right) \quad R = r - y \qquad (2.6.7)$$

The importance of this circuit will become apparent when the simulation of control systems is required. We now turn our attention to the frequency domain and its implications.

2.7 SUMMARY

In this chapter we have introduced the Laplace transform as a mathematical tool for transforming time functions into the frequency domain. We developed the required transforms as required by the control systems environment. Having frequency functions, we required a transform back into the time domain. This was accomplished via the inverse Laplace transform. The two methods used were the partial-fraction expansion and the convolution integral. As physical systems can be shown in differential form, we investigated the solution of differential equations using the Laplace transform. These were differential equations with constant coefficients and zero initial conditions.

We introduced the operational amplifier configuration as a basis for analog simulation. It was found that the gain and integrating blocks were necessary and sufficient for the simulation process. Both blocks provided inversion and gain. The three steps involved in generating a simulation circuit were provided with an emphasis on practical implementation. A summation block configuration was included for future reference.

Problems

2.1. Using the definition of the Laplace transform [(equation (2.2.1)], generate the Laplace transforms for the damped sine and cosine functions.

2.2. Using Table 2.1, find the Laplace transforms for the following time functions.
(a) $f(t) = 6e^{-3t}$　　　　　　　　　　(b) $f(t) = 2 \cos 3t$
(c) $f(t) = 2e^{-6t} \cos 6t$　　　　　　　(d) $f(t) = \cos 3t \cdot \sin 3t$
(e) $f(t) = 3e^{-0.5t} \cos 2t + 4e^{-0.5t} \sin 2t$

2.3. Find the inverse Laplace transforms for the following.
(a) $F(s) = \dfrac{1}{s + 4}$　　　　　　　　(b) $F(s) = \dfrac{3}{2s + 6}$

(c) $F(s) = \dfrac{2}{(s + 1)(s + 2)}$ (d) $F(s) = \dfrac{6}{(s + 2)(s + 3)(s + 4)}$

(e) $F(s) = \dfrac{3}{s(s^2 + s + 4)}$ (f) $F(s) = \dfrac{s^2}{s^2 + 3s + 1}$

2.4. Given the following Laplace transform, show that the convolution integral yields the same time function as the inverse Laplace transform:

$$F(s) = \frac{1}{(s + 3)(s + 5)}$$

2.5. Given the following Laplace transform, show that convolution enjoys the commutative property—that is, $f_1(t) * f_2(t) = f_2(t) * f_1(t)$, where the asterisk indicates the convolution operation.

$$F(s) = \frac{3}{(s + 2)(s + 6)}$$

2.6. Generate a computer program (language of your choice) that will perform the convolution process on the following type of Laplace transform. The program should ask for the number of terms and the time constant for each term. Verify the operation of your program with the Laplace transform given in Problem 2.5.

$$F(s) = \frac{k}{(s + p_1)(s + p_2) \cdots (s + p_n)}$$

2.7. Given the following transfer function.
  (a) What is the steady-state value of the time function?
  (b) If the function is delayed 5 seconds, what is the new Laplace transform?

$$F(s) = \frac{s^3 + 6s^2 + 3s + 4}{s(s^3 + 3s^2 + 2s + 2)}$$

2.8. Modify the program given in Fig. 2.2 such that all roots are obtained. Use your program to help in finding the inverse Laplace transform for the following.

$$F(s) = \frac{3}{s^3 + 3s^2 + 6s + 8}$$

2.9. Consider the circuits given in Fig. P2.9 (a)–(c). Use Laplace transforms to calculate the voltage $v_0(t)$ for all times. Assume the switch (S_1) closes at $t = 0$, $R = 1 \text{ M}\Omega$, $C = 1 \text{ }\mu\text{F}$, $L = 1 \times 10^{-3}$ H, and all initial conditions are equal to zero. Plot all the output voltages with respect to time and comment on the outputs.

2.10. Consider Example 2.15. Find the governing differential equation for the system if $k = 2$, $M = 1$, and $B = 2$. Assume the input $u(t)$ is a unit impulse and that all appropriate units are implied and the system is initially at rest.

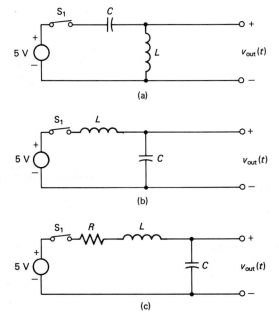

Figure P2.9   Circuits for Problem 2.9

This system is said to be underdamped.   Use Laplace transforms to find the time solution and plot the movement with respect to time.   If $B > 2.83$, the system is considered overdamped. Redo the question if $B = 4$.

2.12. Consider the mechanical system in Fig. P2.12.   Generate the governing differential equation for the system.   If $M = 1$, $B = 2$, $k = 2$, use Laplace transforms to find the solution to the system.   Assume the input is a unit step ($R(s) = 1/s$) and all initial conditions are equal to zero with all constants having appropriate units.   If the springs were of equal length but with different spring constants, how would the system response change?

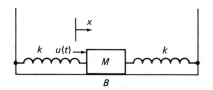

Figure P2.12   Mechanical system for Problem 2.12

2.13. Using ideal operational amplifiers, find the output voltages for the circuits in Fig. P2.13 (a)–(d).

2.14. Generate a circuit diagram to perform the analog simulation for Problem 2.12.   Use Fig. 2.13 as a model.   Build the circuit using the LM324 quad operational amplifier circuitry.   Connect the circuit as indicated in Fig. 2.15 with an input of 1 V.   Verify the operation of your circuit by comparing the results with those obtained mathematically in Problem 2.12.

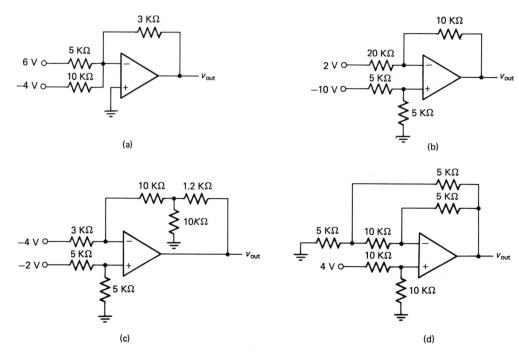

Figure P2.13   Circuits for Problem 2.13

2.15. You are working at Micro Automation, Inc., a company that is considering the building of a control system to position a spacecraft platform. The governing differential equation for the position of the platform ($y(t)$) with respect to the motor shaft position is given by

$$\ddot{y} + c \cdot \dot{y} + 2y = 2 \cdot r(t)$$

where $r(t)$ represents the system input and c is a variable to be decided upon. All initial conditions are assumed to be zero and all units appropriately considered. The variable c can be modified by the control scheme, and you are required to predict the response of the platform, given different values of c, due to a unit step input ($R(s) = 1/s$). This is equivalent to an operator requiring the platform to position itself at one unit of height.
(a) Mathematically predict the system response if $c = 0, 1.4, 3, 5$. Graph each response.
(b) Provide an analog simulation of each case and verify the predictions.
(c) Which response do you feel to be the best for this physical system? Why? What errors do you feel are present with analog similation? What limits in terms of voltage levels exist with the operational amplifier? If the operational amplifier exceeds these levels, it is considered to be in saturation. How does this affect the simulation?

Continuous

3

Transfer Functions

3.1 INTRODUCTION

Chapter 1 introduced the control system as an arrangement of components. In chapter 2 we found that each physical component could be shown as a differential equation and therefore solved via Laplace transforms. Each component consisted of an output generated by the component's operation on the input. If we had some way of showing this operation, then each component would be unique. This operation is in fact the *transfer function* of the component and is the topic of this chapter.

We will investigate the properties of the transfer function and introduce new terms associated with the frequency domain, namely poles and zeroes. We will also explore the *s*-plane and its time-domain implications and study the importance of the characteristic equation.

3.2 DEFINITION OF A TRANSFER FUNCTION

Simply put, a **transfer function** is the ratio of output to input of a given component. Consider the component in Fig. 3.1. If we designate $p(t)$ as the

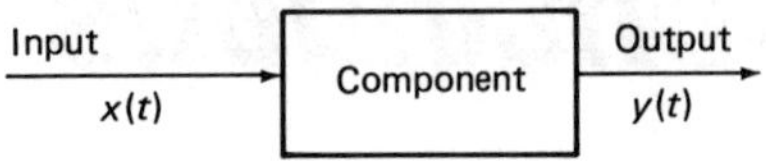

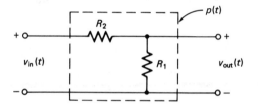

Figure 3.1 Typical component in time

transfer function for this component, we have,

$$p(t) = \frac{y(t)}{x(t)} \tag{3.1.1}$$

EXAMPLE 3.1

Consider the physical component as shown in Fig. 3.2. Find the transfer function $p(t)$.

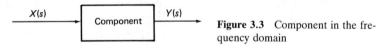

Figure 3.2 Physical component for Example 3.1

SOLUTION   By using circuit analysis we obtain

$$p(t) = \frac{v_{\text{out}}(t)}{v_{\text{in}}(t)} = \frac{R_1}{R_1 + R_2} \tag{1}$$

From equation (1) it is evident that our transfer function defines the operation implied by our component. If we require the system output given a specific input, we simply multiply the input by its transfer function. (Remember that a component could be a very complicated physical system and could contain other components in cascade.)

Figure 3.3 Component in the frequency domain

If a transfer function exists in time, then it must exist in the frequency domain. If we transform the component in Fig. 3.1 to the frequency domain, we obtain the component in Fig. 3.3. If we designate the transfer function as $P(s)$, we obtain

$$P(s) = \frac{Y(s)}{X(s)} \tag{3.1.2}$$

EXAMPLE 3.2

Consider the physical component in the frequency domain as shown in Fig. 3.4. Find the transfer function $P(s)$.

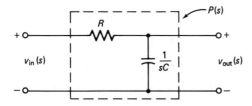

Figure 3.4 Physical component in frequency domain

SOLUTION   Applying circuit analysis, simplifying, we obtain

$$P(s) = \frac{V_{\text{out}}(s)}{V_{\text{in}}(s)} = \frac{\dfrac{1}{RC}}{s + \dfrac{1}{RC}} \tag{1}$$

From equation (1) we can obtain the system output by multiplying the input by the component transfer function. The only difference between this and the previous example is that all functions are in the Laplace domain. If we require a time output for this component, then an inverse Laplace transform is required. To appreciate the implications of the transfer function in the frequency domain, we must investigate its properties.

3.3 PROPERTIES OF THE TRANSFER FUNCTION IN THE LAPLACE DOMAIN

Property 1. *The **transfer function** of a component is the Laplace transform of its unit impulse response.*

Looking at equation (3.1.2), we note that if the input to this system ($X(s)$) is the unit impulse (Dirac delta function), then $X(s) = 1$ and $P(s) = Y(s)$.

Property 2. *A component transfer function is obtained from the component differential equation by applying Laplace transforms and ignoring initial conditions.*

EXAMPLE 3.3

The following differential equation defines the output $y(t)$ with respect to the input $x(t)$ for a specific component. Find the corresponding transfer function in the frequency domain $P(s)$.

$$\ddot{y} + 2\dot{y} + 4y = 2\dot{x} + 4x \qquad (1)$$

SOLUTION   Applying Laplace transforms to equation (1) yields

$$s^2Y(s) + 2sY(s) + 4Y(s) = 2sX(s) + 4X(s)$$

and, solving for the output-to-input relationship, we obtain the required transfer function $P(s)$:

$$P(s) = \frac{Y(s)}{X(s)} = \frac{2s + 4}{s^2 + 2s + 4}$$

Property 3.   *The converse of Property 2 implies that a component differential equation can be generated from the component transfer function by replacing s^n by d^n/dt^n.*

Property 4.   *If we set the numerator of the transfer function to zero and solve for the roots, we have found the component zeroes. If we set the denominator of the transfer function to zero and solve for the roots, we have found the component poles.*

The importance of poles and zeroes will become apparent shortly. For now we will find them strictly as a mathematical exercise.

EXAMPLE 3.4

Using the transfer function ($P(s)$) from Example 3.3, find the component zeroes and poles.

SOLUTION   If we set the numerator to zero, we find one root or zero, which is given by

$$s = -2$$

If we set the denominator to zero, we find two roots or poles, which are given by

$$s_{1,2} = -1 \pm j1.732$$

Property 5.   *The denominator of a transfer function set equal to zero becomes the characteristic equation for the component.*

EXAMPLE 3.5

Find the characteristic equation for the component given in Example 3.3.

SOLUTION   The characteristic equation is generated by setting the denominator of the component transfer function to zero, which in this case gives us

$$s^2 + 2s + 4 = 0$$

Property 6.   *The time response of a component is obtained by taking the inverse Laplace transform of the component transfer function times the required input.   Mathematically this is stated as*

$$y(t) = \mathcal{L}^{-1}[P(s) \cdot X(s)] \tag{3.3.1}$$

EXAMPLE 3.6

If the input to the component of Example 3.3 is a step function of unit height, find the component time response.

SOLUTION   From equation (3.3.1) and the transfer function from example (3.3), we have

$$y(t) = \mathcal{L}^{-1}\left(\frac{1}{s} \cdot \frac{2s + 4}{s^2 + 2s + 4} \right)$$

since $X(s) = 1/s$.   Applying the inverse Laplace transform, we obtain the time response $y(t)$:

$$y(t) = 1 - e^{-t}(\cos 1.732t - 0.577 \sin 1.732t)$$

Property 7.   *The **frequency response** (magnitude and phase) is obtained by replacing s by jω in the component transfer function and putting it in polar form.*

EXAMPLE 3.7

Find the frequency response of the component given in Example 3.3.

SOLUTION   Using the transfer function from Example 3.3 and replacing s by $j\omega$, we have

$$P(j\omega) = \frac{j2\omega + 4}{(j\omega)^2 + j2\omega + 4}$$

Simplifying to

$$P(j\omega) = \frac{4 + j2\omega}{(4 - \omega^2) + j2\omega}$$

and putting into polar form

$$P(j\omega) = \frac{2\sqrt{4 + \omega^2} \measuredangle \arctan\left(\dfrac{\omega}{2}\right)}{\sqrt{(4 - \omega^2)^2 + 4\omega^2} \measuredangle \arctan\left(\dfrac{2\omega}{4 - \omega^2}\right)}$$

we obtain the magnitude and phase, respectively:

$$|P(j\omega)| = \frac{2\sqrt{4 + \omega^2}}{\sqrt{(4 - \omega^2)^2 + 4\omega^2}}$$

$$\measuredangle P(j\omega) = \arctan\left(\frac{\omega}{2}\right) - \arctan\left(\frac{2\omega}{4 - \omega^2}\right)$$

It is left as an exercise for the reader to plot the magnitude and phase with respect to frequency (ω). What kind of filter is exhibited by this physical component? What is the significance of the phase plot? Remember your predictions, as we will revisit the frequency response of a component.

The properties introduced above for transfer functions are very important, and the reader needs to understand them. In Property 4 we found that the Laplace operator can have a real and imaginary part. This implies that the poles and zeroes can be plotted on a cartesian plane. Let's investigate this plane.

3.4 THE S-PLANE

The *s*-plane (or **complex plane**) allows the representation of the poles and zeroes of a component. The real part is given by the Greek letter sigma (σ) and the imaginary part by omega (ω). The imaginary part also represents the frequency component, and for this reason the *s*-plane is also known as the **frequency plane**. Figure 3.5 depicts the *s*-plane.

It was stated earlier that Table 2.1 represented all the Laplace transforms

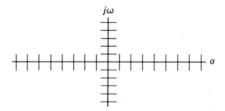

Figure 3.5 The *s*-plane

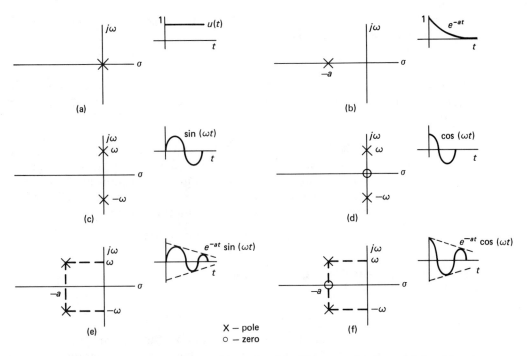

Figure 3.6 Pole-zero patterns: (a) step function, (b) exponential decay, (c) sine, (d) cosine, (e) damped sine, (f) damped cosine

required.   It would be interesting to plot the pole-zero patterns for these entries and at the same time indicate the time functions implied.   Figure 3.6(a)–(f) provides this information.

The reader is asked to predict the time function that has a pole on the positive real axis of the *s*-plane.   The *s*-plane will be important in deciding the *stability* of a system (a term that will be defined in the next chapter), and it is important that the reader be able to visualize the time function given a pole-zero plot.

EXAMPLE 3.8

Given the pole-zero plot for a component shown in Fig. 3.7, indicate the time functions involved in the time response.

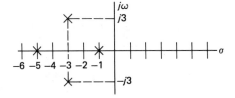

Figure 3.7 Pole-zero pattern for a component

SOLUTION   Using Fig. 3.6, we find the time response depicted in Fig. 3.7 as a function of the following:

$$e^{-t}, \qquad e^{-5t}, \qquad e^{-3t} \sin 3t$$

It is important to note that the actual time response still requires mathematical work, yet the time functions involved in the response are available from the pole-zero plot.

It is evident that the poles of a component are necessary to predict the time response. This is displayed in equation (2.3.3) in terms of the partial-fraction expansion of a component. To obtain the location of the poles we have in fact solved the characteristic equation of the component transfer function. Let's investigate the characteristic equation further.

3.5 THE CHARACTERISTIC EQUATION OF A COMPONENT

The characteristic equation is very important in the design and analysis of control systems. We find the solution to the characteristic equation generating the poles for the component in question. The poles indicate the time functions involved in the component time response. To appreciate the importance of this equation we turn to an example.

EXAMPLE 3.9

A large control system is an accumulation of a large number of small control systems working together. To the large control system, the small one becomes a component. The small control system or component shown in Fig. 3.8 has a variable gain (K) which can be modified to change system response. The control engineer would like to know the change in response as the gain (K) increases from zero.

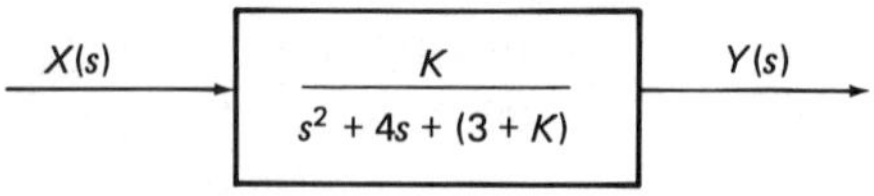

$X(s)$ → $\dfrac{K}{s^2 + 4s + (3 + K)}$ → $Y(s)$

Figure 3.8   Component for Example 3.9

SOLUTION   From Fig. 3.8 we find the characteristic equation to be

$$s^2 + 4s + (3 + K) = 0 \tag{1}$$

If we let $K = 0, 1, 2, 3$ and plot the corresponding pole locations on the

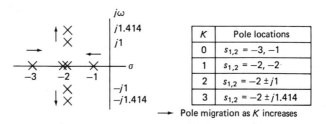

Figure 3.9 shows the following pole locations table:

K	Pole locations
0	$s_{1,2} = -3, -1$
1	$s_{1,2} = -2, -2$
2	$s_{1,2} = -2 \pm j1$
3	$s_{1,2} = -2 \pm j1.414$

→ Pole migration as K increases

Figure 3.9 Pole migration for Example 3.9

s-plane, we can predict the time functions involved in the component response as K varies. Figure 3.9 shows the pole migration as the gain (K) varies.

From Fig. 3.9 we find the time response strictly damped exponential for $K \leq 1$ and a damped sine for $K > 1$. As K increases, we find an increase in the frequency content of our component. This implies a larger bandwidth and therefore greater speed. We arrived at this conclusion in our non-mathematical analysis of a control system in Chapter 1.

3.6 SUMMARY

In this chapter we defined the transfer function in both the time and frequency domains as the ratio of output to input. We investigated the various properties of the transfer function of a component in the frequency domain. The s-plane was introduced with emphasis on the idea of poles and zeroes. The pole-zero patterns for the transforms in Table 2.1 were generated, and the reader was asked to understand the time-function implications for these patterns.

The poles were necessary to predict the time functions involved in the time response. For this reason the denominator of the transfer function set equal to zero became known as the characteristic equation. From this equation it was possible to predict component response or, as in Example 3.9, the response of a control system.

Problems

3.1. Use the definition of a transfer function to calculate the following transfer functions for the circuit in Fig. P3.1:

$$P_1(s) = \frac{V_{\text{out}}(s)}{V_{\text{in}}(s)}, \qquad P_2(s) = \frac{V_{\text{out}}(s)}{I_{\text{in}}(s)}$$

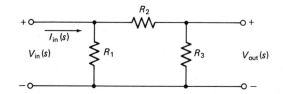

Figure P3.1   Circuit for Problem 3.1

3.2. To measure shaft speed ω_m in control systems a DC tach generator is sometimes used.   The DC tach is coupled to the shaft mechanically, and the
voltage generated is proportional to the speed of the shaft.   This voltage is
equal to the back emf constant K_e times the shaft speed.   The tach generator
has armature resistance R_a and inductance L_a.   A load resistance R_L is put
across the armature to facilitate measurement of the generated voltage.   Figure P3.2 provides the equivalent circuit for this device.   Find the transfer
function of generated voltage $V_{out}(s)$ to shaft speed.   If the shaft is rotating
at a constant speed of 100 radians/sec, $K_e = 1$ volt/radian/sec, and the steady-
state voltage measured is 96 volts, calculate the ratio of load resistance to
armature resistance.

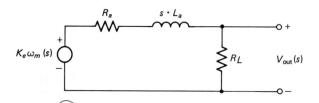

Figure P3.2   DC tach generator circuit
for Problem 3.2

3.3. A system has the following transfer function $P(s)$.   Find the impulse response
in time.

$$P(s) = \frac{s + 6}{(s + 1)(s + 2)(s + 3)}$$

3.4. A system has the following governing differential equation.   Find the system
transfer function and calculate the steady-state output with a unit step input
$(U(s) = 1/s)$.

$$2\ddot{y} + 4\dot{y} + 6y = 5u(t)$$

3.5. Provide an analog simulation for the system with the following transfer function.   Assume the input is a unit step function.

$$P(s) = \frac{6}{s^2 + 4}$$

3.6. Figure P3.6 provides the pole-zero pattern for a system.   If the DC gain is
3, find the system transfer function.

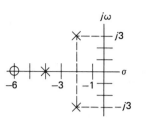

Figure P3.6   Pole-zero pattern for Problem 3.6

3.7. A system has both poles located at -3 on the s-plane. If the following is the characteristic equation for the system, calculate the value of K.

$$s^2 + 6s + (3 + K) = 0$$

3.8. The following transfer function represents a second-order lowpass filter. If $2\zeta_n = 0.2$ and $\omega_n = 1$ rad/sec, plot the magnitude and phase with respect to frequency. Use a computer to calculate the values for magnitude and phase. Be careful with the phase calculations, as computers work between $-\pi/2$ and $\pi/2$ radians.

$$P(s) = \frac{\omega_n^2}{s^2 + 2\zeta_n\omega_n s + \omega_n^2}$$

3.9. For the transfer function in Problem 3.8 show that the maximum value of the frequency response occurs when the frequency ω is equal to ω_r, where

$$\omega_r = \omega_n\sqrt{1 - 2\zeta_n^2}, \qquad \zeta_n < 0.707$$

This frequency is known as the *resonant frequency*. Compare the value attained by this equation and the graphical value in Problem 3.8. Show that the maximum value M_{ω_r} at the resonant frequency is given by

$$M_{\omega_r} = \frac{1}{2\zeta_n\sqrt{1 - \zeta_n^2}}, \qquad \zeta_n < 1$$

Compare this value with your graphical value in Problem 3.8.

3.10. An important quantity that is considered in a frequency response is called *bandwidth* (*BW*). This represents the band of frequencies that is passed by the system. Mathematically it is obtained by calculating the frequency at which the magnitude is equal to 0.707. This is true only for our lowpass filter, since it has a DC gain of unity. Show that the bandwidth of the lowpass filter in Problem 3.8 is given by

$$BW = \omega_n\sqrt{(1 - 2\zeta_n^2) + \sqrt{4\zeta_n^4 - 4\zeta_n^2 + 2}}$$

3.11. Given the circuit in Fig. P3.11, show that the transfer function is given by

$$\frac{V_{out}(s)}{V_{in}(s)} = \frac{1}{s^4(LC)^2 + s^2(3LC) + 1}$$

If $LC = 1$, plot the location of the poles and zeroes. What kind of time response do you expect? Verify your prediction by using the inverse Laplace transform. What is the DC gain of this circuit?

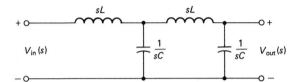

Figure P3.11   Circuit for Problem 3.11

3.12. Consider the bridged-T circuit shown in Fig. P3.12. This circuit is used in AC control systems as a filter network. Show that the transfer function of the network is

$$\frac{V_{out}(s)}{V_{in}(s)} = \frac{s^2(RC)^2 + s(2RC) + 1}{s^2(RC)^2 + s(3RC) + 1}$$

If $RC = 1$, provide the frequency-response plots for the transfer function. What kind of filtering is implied?

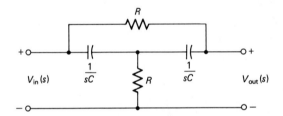

Figure P3.12   Bridged-T circuit for Problem 3.12

3.13. Consider the twin-T circuit as shown in Fig. P3.13. Show that the transfer function for this network is

$$\frac{V_{out}(s)}{V_{in}(s)} = \frac{s^2(RC)^2 + 1}{s^2(RC)^2 + s(4RC) + 1}$$

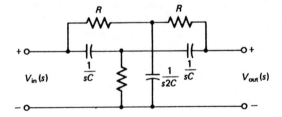

Figure P3.13   Twin-T network for Problem 3.13

If $RC = 1$, provide the frequency-response plots for the transfer function. What kind of filter is this? From this problem and the previous one, can you state the importance of zeroes in the transfer function with reference to the frequency response? A pole-zero plot for both networks will help.

Continuous

4

Stability

4.1 INTRODUCTION

In Chapter 1 we indicated that one disadvantage of closed-loop control systems involved stability. The controls engineer is always concerned about stability in the design of a control system. What constitutes a stable system? How can we tell if a component or control system will be unstable?

In this chapter we will define stability—in time, on the *s*-plane, and in terms of the frequency response. We will also use the Routh table as a mathematical means of calculating stability. It will be evident that the characteristic equation plays a major role in the stability-calculation process.

4.2 DEFINITION OF STABILITY

A system or component can be stable, marginally stable, or unstable.

> *A system or component is **stable** if its impulse response tends to zero as time approaches infinity.*

A system or component is **marginally stable** *if its impulse response is bounded as time approaches infinity.*

A system or component is **unstable** *if its impulse response tends to infinity as time approaches infinity.*

Recall here that the impulse response of a system or component is given by the transfer function. If we are to predict the time response of a system or component, we require the locations of the poles on the *s*-plane. To obtain these locations we have to solve the characteristic equation. Let's look at stability in terms of pole locations.

4.3 STABILITY AND POLE LOCATIONS

If a system is to be stable, then the time response should tend to zero. Referring back to Fig. 3.6, it is evident that these functions have poles in the *left half-plane (LHP)* of the *s*-plane. We can formalize this result in terms of the characteristic equation.

A system or component is **stable** *if the real parts of the roots of the characteristic equation are negative.*

EXAMPLE 4.1

Given the following transfer function for a component, is the component stable?

$$P(s) = \frac{2}{s^3 + 4s^2 + 16s + 32} \tag{1}$$

SOLUTION   From transfer function (1) we obtain the characteristic equation

$$s^3 + 4s^2 + 16s + 32 = 0$$

and, solving for the roots (poles), we obtain

$$s_1 = -2.59$$

$$s_{2,3} = -0.70 \pm j3.44$$

It is evident that the real parts of all three roots are negative and the component is stable.

If a system is marginally stable, then the time response should be bounded. Again referring back to Fig. 3.6, we find these functions on the $j\omega$ axis. That is,

> *A system or component is **marginally stable** if at least one real part of the roots of the characteristic equation is equal to zero.*

EXAMPLE 4.2

You are required to demonstrate a marginally stable system that exhibits a sinusoidal oscillation of 1 rad/sec with a unit step input.

SOLUTION   The transfer function $P(s)$ for this system is given by

$$P(s) = \frac{Y(s)}{X(s)} = \frac{1}{s^2 + 1}$$

and by applying Property 3 of transfer functions we obtain the system differential equation

$$\ddot{y} + y = x(t)$$

A demonstration implies simulation.   From analog simulation we obtain the circuit shown in Fig. 4.1 for the system differential equation.

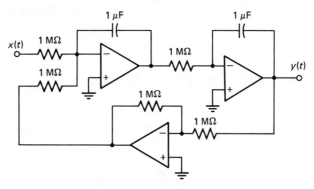

Figure 4.1   Analog simulation of marginally stable system

If a system is unstable, then the time response should tend to infinity. The reader was asked in Chapter 3 to predict the time functions that had poles on the positive real axis.   These functions do in fact tend to infinity as time approaches infinity.   Having poles in the *right half-plane* (*RHP*) of the *s*-plane promotes instability, and the control systems designer should avoid this. In terms of the characteristic equation we have

> *A system or component is **unstable** if at least one real part of the roots of the characteristic equation is positive.*

EXAMPLE 4.3

The transfer function $P(s)$ for a control system is given below. Is the system stable?

$$P(s) = \frac{3}{s^3 + 2s^2 + 5s - 6}$$

SOLUTION   The characteristic equation is given by

$$s^3 + 2s^2 - 5s - 6 = 0$$

and, solving for the roots (poles) (you can use the BASIC program given in Chapter 2), we obtain

$$S_{1,2,3} = -1, -3, 2$$

We have one pole on the RHP, which is sufficient to make the system unstable.

If a component or system exhibits a characteristic equation of a higher order than three, the root-finding method becomes laborious. To accommodate these systems we require a mathematical technique which simplifies the process. That technique is known as the *Routh table.*

4.4 STABILITY AND THE ROUTH TABLE

Consider the general nth-order characteristic equation given by

$$a_n s^n + a_{n-1} s^{n-1} + \cdots + a_1 s + a_0 = 0 \qquad (4.4.1)$$

where a_n, $a_{n-1} \cdots$, a_0 are constant coefficients.

Using the coefficients above, we generate the following **Routh table:**

a_n	a_{n-2}	a_{n-4}		a_1
		$\cdots$		
a_{n-1}	a_{n-3}	a_{n-5}		a_0
b_1	b_2	b_3	$\cdots$	
c_1	c_2	c_3	$\cdots$	

where

$$b_1 = \frac{a_{n-1}a_{n-2} - a_n a_{n-3}}{a_{n-1}}$$

$$b_2 = \frac{a_{n-1}a_{n-4} - a_n a_{n-5}}{a_{n-1}}, \quad \ldots$$

and

$$c_1 = \frac{b_1 a_{n-3} - a_{n-1}b_2}{b_1}$$

$$c_2 = \frac{b_1 a_{n-5} - a_{n-1}b_3}{b_1}, \quad \ldots$$

The process continues until all entries in a row are zero. If the system is stable, we have the following:

> *A system or component is* **stable** *iff all the entries of the first column in the Routh table are of the same sign.*

Mathematically we are ensuring that all real parts of the roots of the characteristic equation are negative. We also obtain some information about the positive real parts:

> *The number of roots with positive real parts is equal to the number of sign changes in the first column of the Routh table.*

EXAMPLE 4.4

Given the following transfer function $P(s)$ for a control system, use the Routh table to investigate its stability:

$$P(s) = \frac{s + 9}{s^4 + 6s^3 + 3s^2 + 4s + 1}$$

SOLUTION   From the transfer function we obtain the characteristic equation:

$$s^4 + 6s^3 + 3s^2 + 4s + 1 = 0$$

and, setting up the Routh table, we obtain

1	3	1
6	4	0
2.33	1	0
1.43	0	
1	0	
0		

From column 1 we find no sign change, and therefore we have a stable system.

EXAMPLE 4.5

The following transfer function for a control system is believed to contain poles in the RHP. The designer would like to know the actual number.

$$P(s) = \frac{s + 3}{s^6 + 4s^5 + 3s^3 + 2s^2 + 6s + 4}$$

SOLUTION From the transfer function we obtain the characteristic equation:

$$s^6 + 4s^5 + 3s^3 + 2s^2 + 6s + 4 = 0$$

and, setting up the Routh table, we obtain

1	0	2	4
4	3	6	0
−0.75	0.5	4	0
5.67	27.3	0	
4.11	4.0	0	
21.8	0		
4.0	0		
0			

C = sign change

It is interesting to note the zero entry due to the absence of the coefficient on the s^4 term. From column 1 we find two sign changes, reflecting the same number of poles in the RHP.

The Routh table can be utilized to investigate the range of certain constants in a characteristic equation to insure the system remains stable. This is of interest to the designer and is investigated in the next example.

EXAMPLE 4.6

A designer generates the following transfer function for a control system. The variable A reflects a physical constant for one of the components used in the system. He would like to know the range of values available for this constant to insure stability.

$$P(s) = \frac{10(s + 3)}{s^3 + (3 + A)s^2 + (3A + 10)s + 40}$$

SOLUTION From the transfer function we obtain the characteristic equation:

$$s^3 + (3 + A)s^2 + (3A + 10)s + 40 = 0$$

and, setting up the Routh table, we obtain.

$$
\begin{array}{|cc}
1 & 3A + 10 \\
3 + A & 40 \\
\dfrac{3A^2 + 13A - 10}{3 + A} & 0 \\
40 & 0 \\
0 &
\end{array}
$$

To insure no sign changes in column 1 of the Routh table, we require the following two conditions:

$$3 + A > 0 \qquad\qquad (1)$$

$$\frac{3A^2 + 19A - 10}{3 + A} > 0 \qquad\qquad (2)$$

From condition (1) we have

$$A > -3$$

From condition (2) we have

$$(A - 0.49) < 0 \quad \text{and} \quad (A + 6.8) < 0$$
$$A < -6.8$$

$$or \qquad (A - 0.49) > 0 \quad \text{and} \quad (A + 6.8) > 0$$
$$A > 0.49$$

We find the constant A must be greater than 0.49 to insure a stable system. The reader is asked to identify the type of stability if $A = 0.49$?

Up until now we have viewed stability in terms of the s-plane and the locations of system poles. It is common practice for the design engineer to use the frequency plots (magnitude and phase) to modify the response of a particular control system. How do we identify a stable system using these plots? Let's find out.

4.5 STABILITY AND FREQUENCY RESPONSE

In Chapter 1 we indicated that our closed-loop control system required negative feedback in order to generate an error. This inversion is equivalent to a phase shift of $-180°$. If our plant, compensation network, and transducer contribute another $-180°$ of phase shift, then our error will increase in time. This will increase the system output and generate an unstable system. In order for the output to grow with time we require a gain of at least unity in our control loop.

In terms of our frequency-response curves we are saying that if there is any frequency component with at least unity gain having a phase shift of at least $-180°$, then we have an unstable system. Mathematically, if our transfer function $P(s)$ is evaluated for a frequency (ω_c) such that the magnitude is unity, we have

$$|P(j\omega_c)| = 1 \qquad (4.5.1)$$

which will generate our crossover frequency (ω_c). If we evaluate the phase shift at this crossover frequency and find it to be greater than $-180°$, then our system is unstable. Notice that the sign on the phase shift represents direction and not magnitude: a phase shift of $-200°$ is greater than a phase shift of $-180°$. Therefore our system is unstable if

$$\sphericalangle P(j\omega_c) \geq -180° \qquad (4.5.2)$$

EXAMPLE 4.7

The following is a transfer function combining the compensation network, plant and transducer. The design engineer requires the maximum gain K in the system if it is to remain stable. The crossover frequency ω_c is also required.

$$P(s) = \frac{K}{s^3 + 3s^2 + 2s + 1} \qquad (1)$$

SOLUTION   Using equation (4.5.1) and transfer function (1), we obtain

$$1 = \frac{K}{\sqrt{1 - 3\omega_c^2)^2 + (2\omega_c - \omega_c^3)^2}} \qquad (2)$$

If the phase shift of transfer function (1) is to be $-180°$, we have

$$-180° = \arctan\left[\frac{2\omega_c - \omega_c^3}{1 - 3\omega_c^2}\right] \qquad (3)$$

Taking tangents of both sides of equation (3) and simplifying, we obtain the crossover frequency (ω_c) to be 1.414 rad/sec. If we substitute this frequency into equation (2), we find the maximum gain K to be 5.

This technique of using frequency-response curves for control system analysis is known as *Bode analysis* and will be covered in detail in Chapter 8. At this time the reader is asked to understand the uses of the frequency-response curves to predict system stability. In Chapter 5 we turn our attention to defining our control system in Fig. 1.1 in the Laplace domain.

4.6 SUMMARY

In this chapter we defined the various forms of stability and indicated that the control system should be a stable environment. We found the transfer function and characteristic equation instrumental in the calculation of the stability of a system or component. It was evident that a stable system would have all its poles in the LHP of the *s*-plane. To handle systems of higher order, the Routh table provided the means by which stability could be calculated.

When the frequency-response curves were analyzed it was found that if a frequency component had at least unity gain with a phase shift of at least $-180°$, then the system was unstable. This phase shift of $-180°$ plus the initial phase shift of $-180°$, due to negative feedback, promoted output growth with time.

Due all But #8

Problems

4.1. "In designing a filtering network, the design engineer sometimes creates an oscillator. The converse is true." Comment on this statement with reference to stability and pole locations on the *s*-plane.

4.2. The human body is the epitome of stable control systems operating in harmony. Consider the speech control system. When it becomes unstable, stuttering occurs. If the following represents the characteristic equation for the speech control system, find the maximum gain *K* before stuttering occurs. Use the Routh table.

$$s^5 + s^4 + (K + 2)s^3 + 4s^2 + K \cdot s + 1 = 0$$

4.3. Use the Routh table to find the number of roots in the RHP for the following characteristic equations.

(a) $s^2 + 6s + 7 = 0$ (b) $s^5 + 6s^3 + 7s + 1 = 0$

(c) $s^3 + s^2 + s + 7 = 0$ (d) $s^4 + 3s^3 + 2s^2 + s + 4 = 0$

(e) $s^{10} + 3s^9 + s^8 + 6s^7 + 3s^5 + 4s^3 + s^2 + 64 = 0$

(f) $s^4 + 10s^3 + 35s^2 + 50s + 66 = 0$

4.4. Consider the simulation circuit as shown in Fig. P4.1. *(See next page.)*

(a) Find the system transfer function.

(b) Is it stable?

(c) Verify your prediction by building and testing the circuit.

4.5. Consider the following transfer function $P(s)$. Provide a pole-zero plot for $P(s)$. Is the system stable?

$$P(s) = \frac{1}{s^2 + 4s + 3}$$

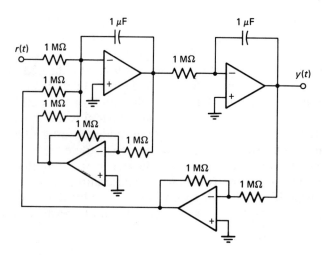

Figure P4.1 Simulation circuit for Problem 4.1

4.6. The following transfer function is similar to that in Problem 4.5 with the addition of a zero. If a pole-zero pattern is generated, it will be evident that the zero lies in the RHP. Is the system still stable?

$$P(s) = \frac{s - 1}{s^2 + 4s + 3}$$

4.7. If each transfer function in Problems 4.5 and 4.6 is put into a separate closed loop with variable gain K, the following respective characteristic equations emerge. We assume that $K > 0$ at all times. Use the Routh table to investigate the range of values of gain for each characteristic equation to insure system stability. Were your predictions in Problems 4.5 and 4.6 correct? Plot the migration of the poles for each characteristic equation for various values of gain K. What is the function of the zero?

$$s^2 + 4s + (3 + K) = 0$$

$$s^2 + (4 + K)s + (3 - K) = 0$$

4.8. Generate a computer program (language of your choice) to perform the Routh-table calculation. The program should ask the user for the coefficients and return with the mechanical workings of the table. It should indicate the stability of the system and, if unstable, the number of poles on the RHP.

4.9. In public address (PA) systems *positive feedback* is taboo. In this case the system becomes unstable and creates a horrifying sound to the human ear. Explain what is happening in terms of frequency response. If we are considering the time domain, how does time delay play a vital role?

4.10. The following is the open-loop transfer function for a control system.
    (a) Use frequency analysis to find the maximum gain K.

(b) What is the crossover frequency?

(c) For the value of K calculated in (a), plot the pole-zero pattern. What is the status of the system in terms of stability?

$$P(s) = \frac{K}{s^4 + 7s^3 + 9s^2 + 17s + 6}$$

4.11. Consider the following transfer function. Use the Routh table to show that this system is unstable. A young control systems technologist plots the pole-zero pattern and indicates that if there is enough gain in a closed loop containing this transfer function, then it will become stable.

$$P(s) = \frac{s + 3}{s^3 + 12s^2 + 29s - 42}$$

The following is the characteristic equation for the closed-loop control system containing the above transfer function. Prove or disprove the comment made by the technologist.

$$s^3 + 12s^2 + (29 + K)s + (3K - 42) = 0$$

Continuous

5

Block Representation of Control Systems

5.1 INTRODUCTION

In previous chapters we have created a base for understanding the component as a transfer function. We mentioned in passing that the control system itself could be shown as a component. If so, all the previous analysis of the component would be applicable to the control system itself.

In this chapter we will define the transfer function of the control system itself. We will introduce the canonical form of a control system and show that all complicated control systems can be represented in this manner. The closed-loop characteristic equation will be highlighted, and systems with multiple inputs will be analyzed via the superposition theorem.

Before we get into the mathematics, let's define the terms associated with the canonical form.

5.2 DEFINITION OF TERMS

In Fig. 5.1 we have the **canonical block** representation of a control system. The reader is asked to compare this to Fig. 1.1.

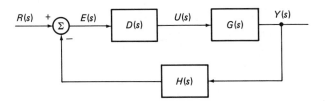

Figure 5.1   Canonical form of control system

From Fig. 5.1 we have the following variables:

Variable	Laplace domain	Time domain
Setpoint	$R(s)$	$r(t)$
Error	$E(s)$	$e(t)$
Control action	$U(s)$	$u(t)$
Output	$Y(s)$	$y(t)$

The component transfer blocks are

Compensation	$D(s)$
Plant	$G(s)$
Feedback	$H(s)$

In examples to follow we use the *special case* of the canonical form. This reduces the complexity of the analysis and in most cases enhances understanding. The special case assumes the transducer is ideal and has a transfer function of unity—that is, $H(s) = 1$. This case is known as **unity feedback**, and Fig. 5.1 is modified as per Fig. 5.2 to reflect the unity-feedback control system.

Having defined all the terms for our canonical control system, we can proceed to find the control system transfer function or the closed-loop transfer function. Since this is a transfer function, all the previous analysis on component transfer functions will be applicable here.

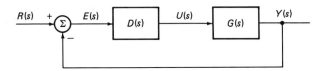

Figure 5.2   Unity-feedback control system

5.3 CLOSED-LOOP TRANSFER FUNCTION FOR CANONICAL FORM

The closed-loop transfer function $P(s)$ is, by the definition of a transfer function, the ratio of output to input, $Y(s)/R(s)$. From Fig. 5.1 the error $E(s)$ is given by

$$E(s) = R(s) - Y(s) \cdot H(s) \qquad (5.3.1)$$

with the control action $U(s)$ being

$$U(s) = E(s) \cdot D(s) \qquad (5.3.2)$$

and the system output $Y(s)$

$$Y(s) = U(s) \cdot G(s) \qquad (5.3.3)$$

If we substitute equation (5.3.1) into (5.3.2) and simplify, we have

$$U(s) = D(s)R(s) - Y(s)H(s)D(s) \qquad (5.3.4)$$

and, substituting equation (5.3.4) into (5.3.3), we obtain the closed-loop transfer function $P(s)$:

$$P(s) = \frac{Y(s)}{R(s)} = \frac{D(s)G(s)}{1 + D(s)G(s)H(s)} \qquad (5.3.5)$$

If we substitute equation (5.3.5) into (5.3.1) and simplify, we obtain an expression for the error:

$$E(s) = \frac{R(s)}{1 + D(s)G(s)H(s)} \qquad (5.3.6)$$

Equation (5.3.6) will be important in finding the steady-state error of our system by applying the final-value theorem. It is also imperative that the control action be available. If we substitute equation (5.3.6) into (5.3.2), we obtain the control action:

$$U(s) = \frac{R(s)D(s)}{1 + D(s)G(s)H(s)} \qquad (5.3.7)$$

From equation (5.3.5) the closed-loop characteristic equation is given by

$$1 + D(s)G(s)H(s) = 0 \qquad (5.3.8)$$

EXAMPLE 5.1

Given the control system shown in Fig. 5.3, assuming unity feedback and dimensionless units, calculate the following:

(a) closed-loop transfer function $P(s)$.
(b) steady-state error $e(\infty)$ for a unit step input ($R(s) = 1/s$).
(c) Value of gain K for a marginally stable system.

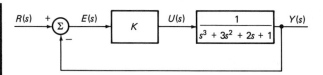

Figure 5.3 Control system for Example 5.1

SOLUTION

(a) From equation (5.3.5) we obtain the closed-loop transfer function:

$$P(s) = \frac{K}{s^3 + 3s^2 + 2s + 1 + K}$$

(b) From equation (5.3.6) we obtain the expression for error:

$$E(s) = \frac{s^3 + 3s^2 + 2s + 1}{s(s^3 + 3s^2 + 2s + 1 + K)}; \qquad R(s) = \frac{1}{s}$$

By applying the final-value theorem we obtain the steady-state error:

$$e(\infty) = \lim_{s \to 0} [s \cdot E(s)] = \frac{1}{1 + K}$$

(c) This is an interesting problem. To be marginally stable at least one of the system poles lies on the imaginary axis of the s-plane. This implies a pole location of the form $s = j\omega_c$, where ω_c is the frequency component. Now the characteristic equation provides the location of all the system poles. This implies that $s = j\omega_c$ is a root of the characteristic equation. If this equation is satisfied, then the value of the gain K should emerge. The characteristic equation is given by

$$s^3 + 3s^2 + 2s + (1 + K) = 0$$

Now substitute $s = j\omega_c$ and simplify:

$$(1 + K - 3\omega_c^2) + j(2\omega_c - \omega_c^3) = 0 \qquad (1)$$

If equation (1) is to be satisfied, then both the real and imaginary parts are equal to zero—that is,

$$1 + K - 3\omega_c^2 = 0 \qquad (2)$$

$$2\omega_c - \omega_c^3 = 0 \qquad (3)$$

From equation (3) we find $\omega_c = 1.414$ rad/sec, and substituting $\omega_c = 1.414$ into equation (2) yields $K = 5$. The reader is asked to revisit Example 4.7 and comment on the resemblance.

We have looked at a very simple control system, and at this time we are unable to simplify a complicated system. Let us look at a system with multiple components and feedback paths.

5.4 BLOCK REDUCTION OF COMPLICATED SYSTEMS

There are many theorems involved in block reductions and the same number to forget. We will see that any complicated system can be reduced by three rules and a systematic approach. The first two rules are new, while the third has been covered.

Blocks in Cascade

Figure 5.4 represents components in cascade and the reduction to one component.

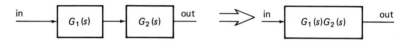

Figure 5.4   Blocks in cascade simplified

Blocks in Parallel

Figure 5.5 represents blocks in parallel and the reduction to one component.

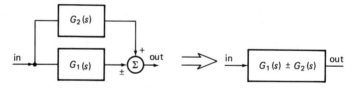

Figure 5.5   Blocks in parallel simplified

Removing a Feedback Block

This essentially means replacing a contained closed-loop control block with its closed-loop transfer function. Figure 5.6 represents the reduction.

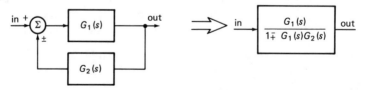

Figure 5.6   Reduction of a feedback block

EXAMPLE 5.2

In high-speed positioning applications of DC motors (servomechanisms) the block diagram in Fig. 5.1 indicates the control scheme. The overall transfer function for this system is required.

SOLUTION   If we study Fig. 5.7, it is evident that the control system is composed of a feedback loop within another feedback loop. This is common in control systems and represents an inner-and-outer-loop architecture.

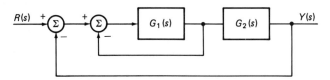

Figure 5.7   Block diagram for servo controller

To simplify, we reduce the inner loop by its closed-loop transfer function—that is,

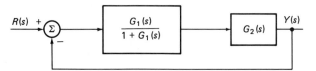

We combine the blocks in cascade:

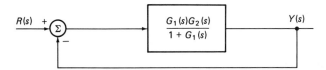

and finally, reducing the closed loop, we obtain the simplified form:

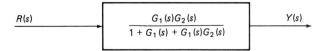

One needs to understand that in a unity feedback system the feedback component is considered to be a block with unity gain. For simplicity it is not shown.

In some cases the simplification cannot be accomplished by use of the previous rules. For these systems we systematically apply the basic definition of the transfer function. Let's investigate with an example.

EXAMPLE 5.3

Consider the control system shown in Fig. 5.8.

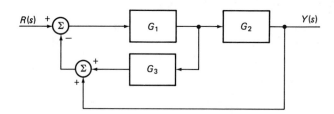

Figure 5.8   Control system for Example 5.3

We see here no application of our basic block reduction methods.   In this case we will apply the basic definition of the transfer function as related to a component.   Figure 5.9 represents the application of the basic transfer function.

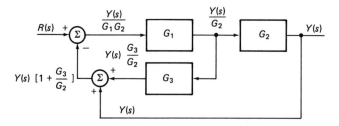

Figure 5.9   System reduction using basics

SOLUTION   From Fig. 5.9 we have at the input summing junction

$$R(s) - Y(s)\left(1 + \frac{G_3}{G_2}\right) = \frac{Y(s)}{G_1 G_2}$$

Simplifying, we obtain the required system transfer function

$$\frac{Y(s)}{R(s)} = \frac{G_1 G_2}{1 + G_1 G_2 + G_1 G_3}$$

This method provides a systematic way of reducing a complicated control system.   It requires knowledge of the definition of a transfer function only.   If the reader feels that our former rules apply, let's redraw Fig. 5.8 as per Fig. 5.10.

It is evident that we have the same system as in Example 5.2.   Therefore the same reduction methods are applicable.   The reader is asked to verify

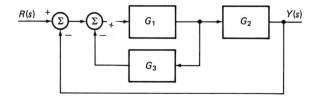

Figure 5.10   Control system in Fig. 5.8 redrawn

that the transfer function from Fig. 5.10 is equal to that generated from Fig. 5.9.

The preceding example exemplifies the fact that basics are very important. If in fact the reader were unable to modify Fig. 5.8 as per Fig. 5.10, the basics would still provide a solution. Let's complete our block reduction by looking at systems with multiple inputs.

5.5 APPLICATION OF THE SUPERPOSITION THEOREM

To find a system response with multiple inputs we can make use of the **superposition theorem**, provided that our system is linear. Let's look at the steps involved.

Step 1   Set all inputs to zero except one chosen input. This will isolate the required input.

Step 2   Calculate the response due to the chosen input acting alone. At this point we can apply our block reduction techniques for complicated systems. It is also important to note that at this point we have generated a transfer function relating the system output with our chosen input.

Step 3   Repeat steps 1 and 2 until each input is considered.

Step 4   The total system response is equal to the algebraic sum of all the individual responses generated in step 2.

EXAMPLE 5.4

A common problem with control systems is the injection of noise (or disturbances) into the closed-loop environment. Figure 5.11 portrays this type of system with disturbance $R_D(s)$. Use the superposition theorem to show that negative feedback decreases the effect of noise, as mentioned among the advantages of negative feedback in Chapter 1.

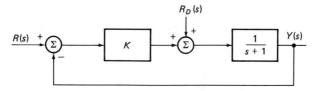

Figure 5.11   Control system with noise input

SOLUTION   By applying step 1, we set our input $R(s) = 0$ and redraw Fig. 5.11 as shown below.   Note the (-1) block to replace the negative feedback.

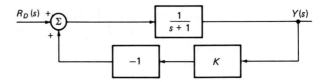

We now generate the transfer function $Y(s)/R_D(s)$:

$$\frac{Y(s)}{R_D(s)} = \frac{1}{s + (1 + K)}$$

If we consider the noise to be random spikes or impulses $(R_D(s) = 1)$, we obtain

$$Y(s) = \frac{1}{s + (1 + K)}$$

and, taking inverse Laplace transforms, we generate the system output due to the random noise:

$$y(t) = e^{-(1+K)t}$$

From the response it is evident that the output due to the noise decays rapidly as the gain (K) in the closed loop increases.

EXAMPLE 5.5

Consider the control system with two inputs $R_1(s)$, $R_2(s)$ as shown in Fig. 5.12.   Find the total system response.

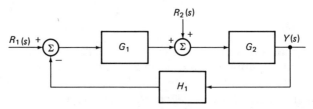

Figure 5.12   Control system for Example 5.5

SOLUTION   To find the response relative to $R_1(s)$ (that is, Y_{R_1}) we let $R_2(s)$ = 0 and redraw Fig. 5.12 as follows:

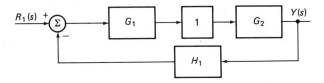

which yields Y_{R_1}:

$$Y_{R_1} = \frac{G_1G_2}{1 + G_1G_2H_1} \cdot R_1(s) \qquad (1)$$

To find the response relative to $R_2(s)$ (that is, Y_{R_2}) we let $R_1(s) = 0$ and redraw Fig. 5.12 as follows:

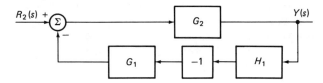

which yields Y_{R_2}:

$$Y_{R_2} = \frac{G_2}{1 + G_1G_2H_1} \cdot R_2(s) \qquad (2)$$

The total system response $Y(s)$ is given by the sum of the individual responses, or equation (1) plus equation (2):

$$Y(s) = \frac{G_2}{1 + G_1G_2H_1} \cdot [G_1R_1(s) + R_2(s)]$$

This concludes the block reduction of complicated control systems. The reader should be confident with this procedure and appreciate the fact that a complicated control system is in fact a combination of very small coupled control systems.

5.6 SUMMARY

In this chapter we defined the canonical form of the control system in the Laplace domain. We introduced the unity feedback control system as a simplified version of the canonical form. We generated the closed-loop transfer function for the canonical form of the control system. We generated an expression for the error and control action for this system. The closed-loop characteristic equation was identified.

We provided three simple rules for block reduction and provided a systematic method using the definition of a transfer function. The superposition theorem was introduced for the solution of control systems with multiple inputs.

Problems

[*Note*: Assume $R(s) = 1/s$ for all steady-state calculations.]

5.1. Comment on the following statement. To understand control systems is to understand components.

5.2. A typical control system is found in Fig. P5.2.
   (a) Find the closed-loop transfer function.
   (b) Find the steady-state error in terms of K.
   (c) Indicate the closed-loop characteristic equation.
   (d) Use the Routh table to find the range of values for K to insure stability.

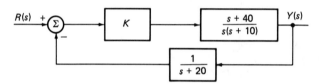

Figure P5.2 Control system for Problem 5.2

5.3. The control system in Fig. P5.3 is to exhibit a 5% steady-state error. What is the value of the gain K? Provide a pole-zero plot at this gain. Find the maximum gain in the system using the Routh table. Take the maximum gain and divide it by the initial gain. This quantity represents the gain margin for this system. Is this quantity useful?

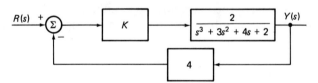

Figure P5.3 Control system for Problem 5.3

5.4. Figure P5.4(a) represents the pole-zero pattern for a plant $G(s)$. The DC gain of the plant is found to be 2. If this plant is put into a closed-loop control system as shown in Fig. P5.4(b), the system poles make a 45° angle with the real axis on the s-plane. What is the value of the gain K? What is the steady-state error?

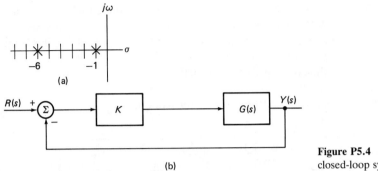

(a)

(b)

Figure P5.4   (a) Pole-zero plot, (b) closed-loop system

5.5. Find the closed-loop transfer function for the control system given in Fig. P5.5.

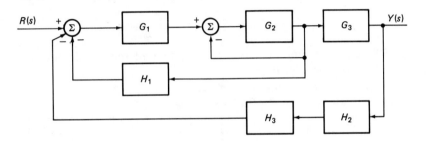

Figure P5.5   Control system for Problem 5.5

5.6. Find the closed-loop transfer function for the control system given in Fig. P5.6.

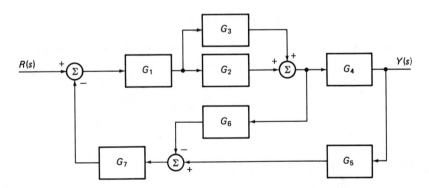

Figure P5.6   Control system for Problem 5.6

5.7. The control system in Fig. P5.7 is typical for the control of one axis in a multiaxis robot.   The inner speed loop is the minor loop, while the outer or major loop is the position loop. The major loop dictates the overall control of the system. In this case we are controlling the position of the axis. The minor loop provides a limit for its quantity controlled. This implies a maximum rate of change of position or speed.

(a) Provide the closed-loop transfer function.

(b) What is the steady-state error?

(c) Find the maximum gain possible in the system.

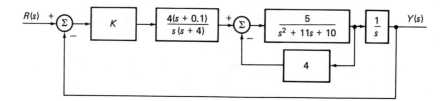

Figure P5.7   Control system for Problem 5.7

5.8. Use the superposition theorem to find the output for the control system shown in Fig. P5.8.

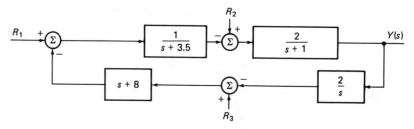

Figure P5.8   Control system for Problem 5.8

5.9. Consider the control system as shown in Fig. P5.9. Show that the system output $Y(s)$ is given by

$$Y(s) = \frac{9(s + 3)}{s^3 + 4s^2 + 9s + 24}$$

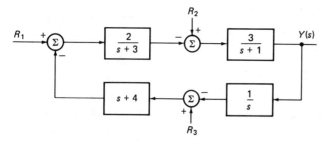

Figure P5.9   Control system for Problem 5.9

If $R_1 = R_2 = R_3 = 1/s$,

(a) Find and plot $y(t)$.

(b) Plot the pole-zero pattern for the system.

(c) How stable is this system?

5.10. In Chapter 2 we developed the method of analog simulation for a component. In this chapter we have stated that a component can also be a control system. It would appear that we could in fact simulate a control system. Consider the transfer function for a DC motor given below.

$$G(s) = \frac{1}{s^2 + 4s + 3}$$

This represents the ratio of shaft speed to armature voltage. If we would like to control the speed of this motor, we could use the basic closed-loop control system as shown in Fig. P5.10(a).

From the figure it is evident that we can simulate $G(s)$ but have yet to generate the summing and gain block to close the loop. If we refer to Fig. 2.8(a) we find a gain block. The gain is given by R_2/R_1 and it is inverting. From Fig. 2.14(b) we find a summing block that inverts one of the inputs. Now Fig. P5.10(a) can be shown as Fig. P5.10(b).

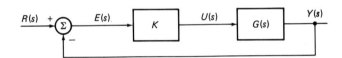

Figure P5.10(a)   Control system for Problem 5.10

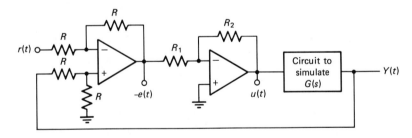

Figure P5.10(b)   Simulation of control system for Problem 5.10

If you compare Fig. 2.14(b) with Fig. P5.10(b), you will find that the quantities $r(t)$ and $y(t)$ are reversed. Why?

Refer to Fig. P.5.10(a).

(a) Find the closed-loop transfer function.

(b) If $R(s) = 1/s$, calculate and plot $e(t)$, $u(t)$, $y(t)$ for $K = 1$ and 2.

Refer to Fig. P5.10(b).

(a) Generate the analog simulation of $G(s)$.

(b) Use $R = 100$ KΩ, $R_1 = 10$ KΩ, $R_2 = 10$ KΩ, 20 KΩ and complete the analog simulation circuit for the control system.

(c) Using a step input of 1 V, obtain the responses of $e(t)$, $u(t)$, $y(t)$ for both values of K.

(d) Compare these with the previous predicted plots.

(e) Investigate higher gains and the effect of saturation in the operational amplifiers.

(f) It appears that our system has less error with higher gains. What is the tradeoff for this benefit?

Part I

Continuous

6

Second-Order Systems

6.1 INTRODUCTION

Why the emphasis on second-order systems? It is evident that most physical plants exhibit a differential equation of order greater than two, and it would seem a futile exercise to study this system at any depth. The reason is quite simple: mathematically we can disassemble any higher-order system into a series of first- and second-order systems. The first-order system provides a decaying contribution, while the second-order system can provide the ringing and overshoot. This implies that if we fully understand first- and second-order systems, then we actually understand higher-order systems. Sometimes it is also possible to approximate higher-order systems with second-order systems.

   In this chapter we will formally define the second-order plant and investigate its step response. We will also derive transfer functions for some well-known second-order physical plants.

6.2 FORMAL DEFINITION OF A SECOND-ORDER SYSTEM

Any plant which exhibits the linear constant-coefficient second-order differential equation of the form

$$\frac{d^2y}{dt^2} + 2\zeta_n\omega_n \frac{dy}{dt} + \omega_n^2 y(t) = \omega_n^2 u(t) \qquad (6.2.1)$$

is referred to as a second-order plant. The constant ζ_n is called the **damping ratio** and the constant ω_n is called the **undamped natural frequency**. Taking Laplace transforms of (6.2.1), with zero initial conditions, we have

$$s^2 Y(s) + 2\zeta_n\omega_n s Y(s) + \omega_n^2 Y(s) = \omega_n^2 U(s) \qquad (6.2.2)$$

and, simplifying, we obtain the second-order plant transfer function

$$G(s) = \frac{Y(s)}{U(s)} = \frac{\omega_n^2}{s^2 + 2\zeta_n\omega_n s + \omega_n^2} \qquad (6.2.3)$$

The plant characteristic equation is given by

$$s^2 + 2\zeta_n\omega_n s + \omega_n^2 = 0 \qquad (6.2.4)$$

At this point it is very important to understand the difference between the system and plant characteristic equations. The latter is simply taken from the plant itself, while the former is taken from the closed-loop transfer function. If the system characteristic equation happens to retain a second-order form, then the subscripts will be changed to indicate the modified form. The form of the second-order system characteristic equation will be

$$s^2 + 2\zeta_m\omega_m s + \omega_m^2 = 0 \qquad (6.2.5)$$

where ω_m and ζ_m could be functions of ω_n and ζ_n, respectively. As before, ω_m represents the system undamped natural frequency and ζ_m represents the system damping ratio. These parameters will reflect the changes in the original plant undamped natural frequency ω_n and plant damping ratio ζ_n. Consider the following example.

EXAMPLE 6.1

Given the following control system:

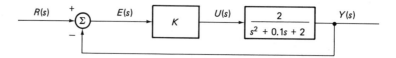

The plant characteristic equation is

$$s^2 + 0.1s + 2 = 0$$

and comparison with (6.2.4) yields

$$2\zeta_n\omega_n = 0.1, \qquad \omega_n^2 = 2$$

The system transfer function is

$$\frac{Y(s)}{R(s)} = \frac{2K}{s^2 + 0.1s + (2 + 2K)}$$

with the system characteristic equation being

$$s^2 + 0.1s + (2 + 2K) = 0$$

and comparison with (6.2.5) yields

$$2\zeta_m\omega_m = 0.1, \qquad \omega_m^2 = 2 + 2K$$

It is evident that the system undamped natural frequency is increased by the factor $2K$ as compared to the plant undamped natural frequency. Since the system undamped natural frequency is increasing, the system damping ratio is decreasing as compared to the plant damping ratio. This change of parameters will promote a specific system response, as we will see in the next section.

6.3 DAMPING RATIO

To understand the effect of changing the plant damping ratio it is useful to track the plant poles for various values of this constant. Before we look at the five cases, we have to find the locations of the plant poles in terms of the plant damping ratio. The plant poles are found by finding the roots of the plant characteristic equation. Taking (6.2.4), finding the roots, and simplifying, we have

$$s_{1,2} = -\zeta_n\omega_n \pm \omega_n \sqrt{\zeta_n^2 - 1} \tag{6.3.1}$$

Case 1: $\zeta_n > 1$

From (6.3.1) it is evident that both roots are real and the response of this system will be strictly exponential or an **overdamped system**. Figure 6.1 provides the locations of the poles.

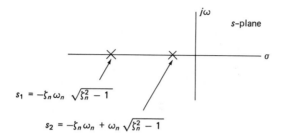

Figure 6.1   Locations of poles for $\zeta_n > 1$

Case 2: $\zeta_n = 1$

In this case equation (6.3.1) becomes

$$s_{1,2} = -\zeta_n \omega_n \qquad (6.3.2)$$

From (6.3.2) it is evident that the roots are real and equal. The response will still be exponential, and this special case is **critically damped**. Figure 6.2 provides the locations of the poles.

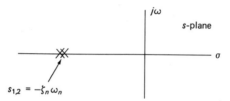

Figure 6.2   Locations of poles for $\zeta_n = 1$

Case 3: $0 < \zeta_n < 1$

From (6.3.1) the locations of the poles are

$$s_{1,2} = -\zeta_n \omega_n \pm j\omega_n \sqrt{1 - \zeta_n^2} \qquad (6.3.3)$$

It is evident that the pole locations are complex conjugates, which implies a decaying sine or cosine response. This case represents an **underdamped plant**. Figure 6.3 provides the locations of the poles.

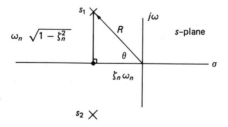

Figure 6.3   Locations of poles for $0 < \zeta_n < 1$

This case is of special interest, as most control systems will operate in this area. In Fig. 6.3 we have indicated a distance from the origin to the pole as the quantity R and the angle this vector makes with the real axis as θ. By the Pythagorean theorem we have

$$R = \omega_n \tag{6.3.4}$$

and simple trigonometry yields

$$\cos \theta = \zeta_n \tag{6.3.5}$$

From (6.3.4) we find the plant undamped natural frequency as the length of the vector. The cosine of the angle θ yields the plant damping ratio, as seen in (6.3.5). The plant has a frequency component given by the plant **damped natural frequency** (ω_d):

$$\omega_d = \omega_n \sqrt{1 - \zeta_n^2} \tag{6.3.6}$$

From (6.3.6) it is evident that the plant will have a frequency component less than the undamped natural frequency and will be equal only if the angle θ approaches 90°. This brings us to the next case.

Case 4: $\zeta_n = 0$

From (6.3.1) the locations of the poles are

$$s_{1,2} = \pm j\omega_n \tag{6.3.7}$$

In this case the system poles lie on the imaginary axis, which implies an oscillator with a frequency equal to the system undamped natural frequency. Figure 6.4 provides the locations of the poles.

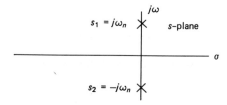

Figure 6.4   Locations of poles for $\zeta_n = 0$

Case 5: $\zeta_n < 0$

From (6.3.1) the locations of the poles are

$$
\begin{aligned}
s_{1,2} &= \zeta_n \omega_n \pm j\omega_n \sqrt{1 - \zeta_n^2}, && -1 < \zeta_n < 0 \\
s_{1,2} &= \zeta_n \omega_n \pm \omega_n \sqrt{\zeta_n^2 - 1}, && \zeta_n \leq -1
\end{aligned}
\tag{6.3.8}
$$

From (6.3.9) it is obvious that the poles cross into the right half of the s-plane and therefore promote an unstable system. Since this type of plant is not naturally found and we would never design a system to these specifications, we will no longer pursue this case.

6.4 STEP RESPONSE

In the previous section we investigated the migration of the plant poles as the plant damping ratio was modified. We predicted the time response based on the location of the poles. To appreciate these predictions we will take a second-order plant and calculate the step response for each case of damping ratio covered in the previous section. This will provide a comparison of plant speed as compared to damping ratio.

EXAMPLE 6.2

Consider the following second-order plant:

$$G(s) = \frac{Y(s)}{U(s)} = \frac{1}{s^2 + 2\zeta_n s + 1} \tag{1}$$

Case 1: $\zeta_n = 2$. Transfer function (1) becomes

$$\frac{Y(s)}{U(s)} = \frac{1}{s^2 + 4s + 1}$$

with the step response being ($U(s) = 1/s$)

$$Y(s) = \frac{1}{s(s^2 + 4s + 1)}$$

and using partial fraction expansion, inverse Laplace transforms, simplifying, we obtain the time response:

$$y(t) = 1 - 1.0774e^{-0.2679t} + 0.0774e^{-3.7321t} \tag{2}$$

Case 2: $\zeta_n = 1$. Transfer function (1) becomes

$$\frac{Y(s)}{U(s)} = \frac{1}{s^2 + 2s + 1}$$

with the step time response

$$y(t) = 1 - e^{-t} - te^{-t} \tag{3}$$

Case 3: $\zeta_n = 0.5$.   Transfer function (1) becomes

$$\frac{Y(s)}{U(s)} = \frac{1}{s^2 + s + 1}$$

with the step time response

$$y(t) = 1 - e^{-0.5t}\cos 0.866t - 0.5774e^{-0.5t}\sin 0.866t \qquad (4)$$

Case 4: $\zeta_n = 0$.   Transfer function (1) becomes

$$\frac{Y(s)}{U(s)} = \frac{1}{s^2 + 1}$$

with the step time response

$$y(t) = 1 - \cos t \qquad (5)$$

Figure 6.5 provides a graph of equations (2), (3), (4), and (5), and it is evident that the speed of response increases as damping ratio decreases. When the damping ratio is equal to zero, we find the response oscillating at the plant undamped natural frequency, as expected.

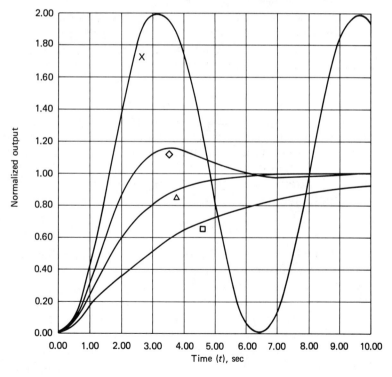

□ $\zeta_n = 2.0$ △ $\zeta_n = 1.0$ ◇ $\zeta_n = 0.5$ X $\zeta_n = 0.0$

Figure 6.5   Step response with varying damping ratio (ζ_n)

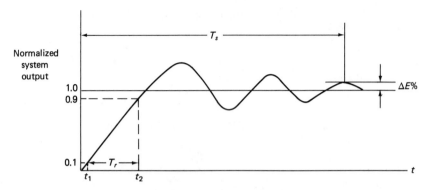

Figure 6.6 System rise time and settling time

Most physical plants appear with a damping ratio which is greater than unity, and once they are put into a control system the system damping ratio can be modified with the existing compensating network. We have been talking about a fast response and yet have no way of indicating this entity. Figure 6.6 defines two new terms to provide us with an intelligent way of comparing system responses. The **rise time** (T_r) and **settling time** (T_s) are valid variables for comparison and for system response specifications. From Fig. 6.6 the system risetime is the time taken for the response to go from 10% to 90% of its final value. Mathematically this is

$$T_r = t_2 - t_1 \tag{6.4.1}$$

which is conveniently obtained from the response graph. It is left as an exercise for the reader to write a computer program to calculate the system rise time, given any system output equation in time.

We will define our system or plant settling time on the assumption that the relative damping ratio is between zero and one. If this is the case, then the output equation, using (6.3.3), in time will be

$$y(t) = 1 - e^{-\zeta_n \omega_n t} \left[K_1 \cos \omega_d t + K_2 \sin \omega_d t \right] \tag{6.4.2}$$

where K_1 and K_2 are the appropriate constants generated from partial-fraction expansion and inverse Laplace transforms with a step input. From (6.4.2) it is evident that the exponential term is responsible for the decaying process. From Figure 6.6 we have defined the sampling time as the time taken for the output to be within a specified bound of the steady-state value. This quantity is shown as $\Delta E\%$ and is specified as a percentage of the steady-state value. This implies that the exponential portion of our response should reach this value if the time is equal to the settling time. Mathematically this is

$$e^{-\zeta_n \omega_n T_s} = \frac{\Delta E\%}{100} \tag{6.4.3}$$

and, solving for the settling time,

$$T_s = \frac{-\ln (0.01 \ \Delta E\%)}{\zeta_n \omega_n} \qquad (6.4.4)$$

From (6.4.4) it is evident that the system settling time can be decreased if either the undamped natural frequency or damping ratio is increased. It should be remembered that equation (6.4.4) relies on the fact that the damping ratio is between zero and one. Let's look at two examples.

EXAMPLE 6.3

Consider the following plant.

$$G(s) = \frac{Y(s)}{U(s)} = \frac{1}{s^2 + 0.2s + 1}$$

Find the plant settling time if it is to settle within 3% of its final value, given a step input.

SOLUTION   The plant characteristic equation is

$$s^2 + 0.2s + 1 = 0 \qquad (1)$$

Comparing (1) with (6.2.4), we have

$$\zeta_n \omega_n = 0.1$$

Substituting the values into (6.4.4), we have

$$T_s = \frac{-\ln (0.01 \times 3)}{0.1}$$

$$= 35 \text{ seconds}$$

EXAMPLE 6.4

Given the following control system:

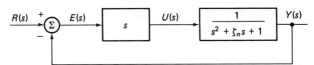

If the system is to settle within 3% of its final value in 4 seconds, then find the original plant damping ratio (ζ_n).

SOLUTION   The closed-loop transfer function is

$$\frac{Y(s)}{R(s)} = \frac{s}{s^2 + s(\zeta_n + 1) + 1} \qquad (1)$$

with the system characteristic equation being

$$s^2 + s(\zeta_n + 1) + 1 = 0 \tag{2}$$

Comparing (2) with (6.2.5), we have

$$\zeta_m \omega_m = \frac{\zeta_n + 1}{2}$$

Since we are dealing with a system, equation (6.4.4) is modified to reflect this. That is,

$$T_s = \frac{-\ln (0.01 \ \Delta E\%)}{\zeta_m \omega_m} \tag{3}$$

Substituting the information into (3), we have

$$4 = \frac{-\ln (0.01 \times 3)}{\dfrac{\zeta_n + 1}{2}}$$

Solving for the plant damping ratio, we have

$$\zeta_n = 0.75$$

This example will help the reader understand the reason for introducing the new subscripts. They provide a method of keeping the original plant un-damped natural frequency and damping ratio as separate entities from the new system undamped natural frequency and damping ratio. This idea will be used throughout this text, and it is imperative that the reader understand the implications.

   As was stated before, the case with the damping ratio between zero and one is of special interest. We will now generate a family of response curves based on the damping ratio varying from 0.1 to 0.9. Since these curves are normalized, the reader can use the same for future comparison purposes. From (6.2.3) the step response ($U(s) = 1/s$) is

$$Y(s) = \frac{\omega_n^2}{s(s^2 + 2\zeta_n \omega_n s + \omega_n^2)} \tag{6.4.5}$$

Using partial-fraction expansion, we have

$$Y(s) = \frac{A}{s} + \frac{B_s + C}{(s + \zeta_n \omega_n)^2 + \omega_n^2(1 - \zeta_n^2)} \tag{6.4.6}$$

and, solving for the constants A, B, C, we have

$$Y(s) = \frac{1}{s} - \left[\frac{s + \zeta_n \omega_n}{(s + \zeta_n \omega_n)^2 + \omega_n^2(1 - \zeta_n^2)} \right.$$
$$\left. + \frac{\zeta_n}{\sqrt{1 - \zeta_n^2}} \frac{\omega_n \sqrt{1 - \zeta_n^2}}{(s + \zeta_n \omega_n)^2 + \omega_n^2(1 - \zeta_n^2)} \right] \qquad (6.4.7)$$

Taking inverse Laplace transforms, we generate the general form of the time response:

$$y(t) = 1 - e^{-\zeta_n \omega_n t} \left[\cos\left(\sqrt{1 - \zeta_n^2}\, \omega_n t\right) + \frac{\zeta_n}{\sqrt{1 - \zeta_n^2}} \sin\left(\sqrt{1 - \zeta_n^2}\, \omega_n t\right) \right] \qquad (6.4.8)$$

If we vary time ($t = 1/\omega_n$) from 0 to 10 seconds for each case of damping ratio (0.1 to 0.9), then a family of response curves will emerge. Figure 6.7 provides the family of response curves.

EXAMPLE 6.5

You are being considered by the Rotation Motor Company for a job opening in their variable-speed drives department. The person from the company provides you with the graph as per Fig. 6.7 and asks the following. "A motor exhibits a 0.1 damping ratio and an undamped natural frequency of 10 rad/sec. With a step input, the motor levels out to a speed of 540 RPM. What are the rise time and maximum speed for the motor?" Your answer will greatly influence your chances of a permanent position with the company.

SOLUTION You indicate that the scales are normalized and proceed to estimate the expected rise time and maximum speed. From Fig. 6.7 it is evident that the curve representing a 0.1 damping ratio has a maximum value and rise time of 1.73 and 1.0 units, respectively. The vertical scale is multiplied by 540 and the horizontal scale is divided by 10. This allows the graph to represent the motor response. Therefore the maximum speed will be 934 RPM and the rise time 0.1 second.

Although this example seems trivial and nonmathematical, it allows a quick estimation of the system response without a lengthy calculation. It also emphasizes the fact that a normalized scheme has an infinite amount of solutions embedded. For a unique solution to a particular problem only the scaling has to be modified.

To appreciate some of the physical plants which exhibit a second-order response, we now turn our focus to this area.

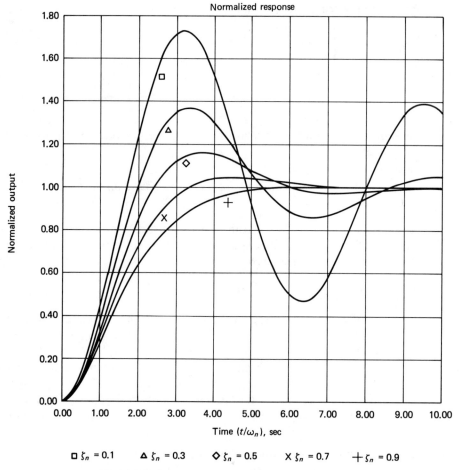

Figure 6.7   System response with varying damping ratio

6.5 EXAMPLES OF SECOND-ORDER SYSTEMS

DC Motor

This device is used to convert electrical energy to mechanical or rotational energy. The motor in question has a separately excited field. If we break this device into its electrical and mechanical components, we should be able to define it mathematically. Since the armature voltage is proportional to

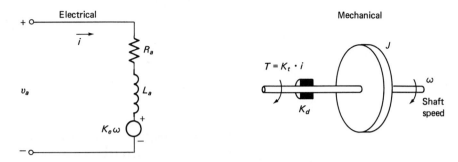

Figure 6.8 Electrical and mechanical sequivalents of a DC motor

speed and the current to torque, we can show the DC motor as in Fig. 6.8, where

v_a = armature voltage   [volts]
ω = motor shaft speed   [rad/sec]
R_a = armature resistance   [ohms]
L_a = armature inductance   [henries]
i = armature current   [amps]
K_e = back EMF constant   [volts/(rad/sec)]
K_t = torque constant   [oz-in./amp]
K_d = viscous damping constant [oz-in./(rad/sec)]
J = total inertia of armature and load   [oz-in.-sec-sec]

From Fig. 6.8 the equivalent voltage equation is

$$v_a = i \cdot R_a + L_a \frac{di}{dt} + K_e \cdot \omega \qquad (6.5.1)$$

with the torque equation being

$$K_t \cdot i = J \frac{d\omega}{dt} + K_d \cdot \omega \qquad (6.5.2)$$

If we take the Laplace transforms of equations (6.5.1) and (6.5.2) with zero initial conditions, we have

$$V_a(s) = I(s) + sI(s)L_a + K_e\Omega(s) \qquad (6.5.3)$$

$$K_t I(s) = sJ\Omega(s) + K_d\Omega(s) \qquad (6.5.4)$$

Substituting equation (6.5.4) into (6.5.3), we obtain

$$\frac{\Omega(s)}{V_a(s)} = \frac{K_t}{L_a J s^2 + s(JR_a + K_d L_a) + K_d R_a + K_e K_t} \qquad (6.5.5)$$

and, simplifying,

$$\frac{\Omega(s)}{V_a(s)} = \frac{\dfrac{K_t}{L_a J}}{s^2 + s\left(\dfrac{R_a}{L_a} + \dfrac{K_d}{J}\right) + \dfrac{K_d R_a + K_e K_t}{L_a J}} \tag{6.5.6}$$

From (6.5.6) we can see that for a particular motor the only parameter that can change is the total inertia (J). This parameter contains the inertia of the armature and load. It is evident that if the load inertia increases, the motor undamped natural frequency will decrease. This implies a larger rise time or a slower step response. The reader is encouraged to continue this type of analysis of the step response with reference to the balance of the parameters.

Simple Suspension System

The system illustrated in Fig. 6.9 is a good approximation for both a suspension and printer head mechanism. The system involves a mass m, spring (spring constant K), and damper (damping constant B). With an applied force ($u(t)$), the mass will be accelerated, with the spring applying a force proportional to distance travelled and the damper a force proportional to the speed of the mechanism.

Using simple laws of physics, we generate the plant differential equation:

$$m\frac{d^2 y}{dt^2} + B\frac{dy}{dt} + Ky = u(t) \tag{6.5.7}$$

Taking Laplace transforms of (6.5.7), with zero initial conditions, we have

$$ms^2 Y(s) + BsY(s) + KY(s) = U(s) \tag{6.5.8}$$

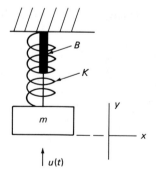

Figure 6.9 Mass-spring-damper mechanism

and, simplifying,

$$\frac{Y(s)}{U(s)} = \frac{\dfrac{1}{m}}{s^2 + s\dfrac{B}{m} + \dfrac{K}{m}} \tag{6.5.9}$$

From (6.5.9) it is evident that our system would oscillate if the damper mechanism disappeared ($B = 0$). What would be the effect on system response if either the mass or spring constant were increased? How does the mass in this system compare with the inertia of the DC motor?

RLC Circuit

In Fig. 6.10 we see a typical RLC circuit. We are concerned with looking at the voltage across the capacitor (v_c) as a function of input voltage (v_{in}).

In Fig. 6.11 we have the equivalent circuit in the Laplace domain.

From Fig. 6.11 the current $I(s)$ is

$$I(s) = \frac{V_{in}(s)}{R + sL + \dfrac{1}{sC}} \tag{6.5.10}$$

The voltage across the capacitor $V_c(s)$ is

$$V_c(s) = \frac{1}{sC} \cdot I(s) \tag{6.5.11}$$

and, substituting (6.5.10) into (6.5.11), we obtain the required transfer function:

$$\frac{V_c(s)}{V_{in}(s)} = \frac{\dfrac{1}{LC}}{s^2 + s\dfrac{R}{L} + \dfrac{1}{LC}} \tag{6.5.12}$$

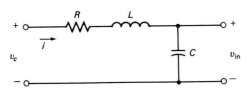

Figure 6.10 RLC circuit

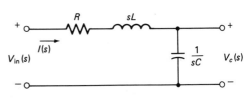

Figure 6.11   Equivalent Laplace circuit

From (6.5.12) we can see the similarity of the resistance to the damper mechanism in the previous example. What is the role of the capacitance and inductance in this circuit in terms of a step response?

Other Second-Order Systems

Another plant which exhibits a second-order nature is a two-tank level system. Since this plant is covered in Example 7.4, it is only mentioned in passing at this point.

A pneumatic valve also exhibits a second-order nature—and the list goes on and on. The main objective is to understand the second-order plant in terms of the undamped natural frequency and damping ratio. With this in place, the reader will have no problem in understanding any physical system with a second-order nature.

6.6 SUMMARY

In this chapter we formally defined the second-order plant in terms of the undamped natural frequency and damping ratio. We tracked the poles on the s-plane as the plant damping ratio was changed. This provided some insight as to the time response expected for the relevant damping ratio. We found it necessary to introduce new subscripts for the undamped natural frequency and damping ratio. This was to accommodate a closed-loop characteristic equation which was of a second order.

The case with the damping ratio being between zero and one was of special interest. Here we introduced the damped natural frequency and a family of normalized response curves for comparison purposes. We provided some examples of physical second-order plants. The emphasis was on how different the systems can be, yet how similar they appear in the Laplace domain. This similarity is a very powerful tool in the development of simulation systems and should be taken seriously by the reader.

Problems

6.1. If you refer back to Problem 3.8, it is evident that our second-order plant is in fact a lowpass filter. Problem 3.10 provides an expression for bandwidth with respect to the damping ratio. Plot the quantity BW/ω_n with respect to damping ratio.   Let the damping ratio vary from 0 to 2.   What is the relationship between bandwidth and damping ratio?

6.2. If the system undamped natural frequency increases and the system damping ratio decreases, what is the effect on settling time?   What about rise time?

6.3. Given the following second-order plant characteristic equation, what is the undamped natural frequency?   What is the damping ratio?

$$2s^2 + 4s + 6 = 0$$

6.4. The control system in Fig. P6.4 is to be critically damped.   What value of gain (K) is required?

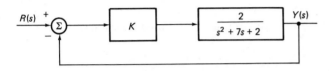

Figure P6.4   Control system for Problem 6.4

6.5. Consider the following nameplate data for a high speed DC motor which is to used in a robotic application.

$$
\begin{aligned}
K_t &= 15.29 &&\text{oz-in./amp} \\
K_e &= 11.30 &&\text{volts/1000 RPM} \\
R_a &= 0.73 &&\text{ohms} \\
L_a &= 100 &&\text{microhenries} \\
K_d &= 1.65 &&\text{oz-in./1000 RPM} \\
J &= 0.017 &&\text{oz-in.-sec}^2 \text{ (motor only)}
\end{aligned}
$$

Show that the shaft-speed to armature-voltage transfer function is given by

$$G(s) = \frac{\Omega(s)}{V_a(s)} = \frac{8.99 \times 10^6}{s^2 + 7300s + 9.77 \times 10^5}$$

The motor is put into the control system as shown in Fig. P6.5.   Calculate the value of gain K required for the control system to exhibit a 0.707 damping ratio.   What are the undamped natural frequency, bandwidth, and rise time for the control system?   What is the amount of overshoot, and how does it compare to the predictions from the normalized curves?   Assume the input is $R(s) = 6.28/s$.   What is the steady-state speed in RPM?   What is the steady-state error?

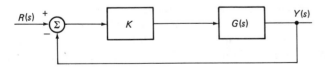

Figure P6.5   Control system for Problem 6.5

6.6. The motor in Problem 6.5 has a load connected to it with an inertia of 0.023 oz-in. sec^2.   This implies a new $G(s)$.   If the control system in Fig. P6.5 is used with the same gain K, find the system damping ratio. What is the new steady-state error? In practical systems the load changes at all times. Would this control scheme be attractive as a speed controller?

6.7. The following transfer function represents a closed-loop transfer function. It is evident that this system is a third-order system.
   (a) If $R(s) = 1/s$, find and graph $y(t)$.
   (b) From Fig. 6.5 estimate the system damping ratio.
   (c) Calculate the damping ratio of the embedded second-order system. How does it compare to that of (b)?
   (d) Plot the pole-zero pattern for the system. It is evident that the embedded second-order poles are dominant, as they contain an exponential decay with a large time constant. What can be said about the response of a higher-order system as compared to the dominant system poles?

$$P(s) = \frac{Y(s)}{R(s)} = \frac{25}{s^3 + 7s^2 + 14s + 20}$$

6.8. Consider the closed-loop transfer function given below.
   (a) Provide a pole-zero plot for the control system.
   (b) What is the dominant system?
   (c) If $R(s) = 1/s$, find and graph $y(t)$.
   (d) Find the damping ratio of the embedded second-order system. Does the statement in Problem 6.7(d) still hold true?

$$P(s) = \frac{Y(s)}{R(s)} = \frac{3}{s^3 + 4.5s^2 + 6s + 2}$$

6.9. Show that a second-order transfer function as given by equation (6.2.3) has a maximum percent overshoot M_p given by

$$M_p = 100e^{-\zeta_n \omega_n \cdot T_p}$$

for a unit step input, with T_p being the time that the maximum overshoot occurs and given by

$$T_p = \frac{\pi}{\omega_n \sqrt{1 - \zeta_n^2}}$$

This is true only for a damping ratio between 0 and 1, and equation (6.4.8) is very helpful.

6.10. Plot the maximum percent overshoot M_p versus damping ratio for the second-order system. Plot T_p versus damping ratio. If possible, use a computer to obtain an accurate plot in each case. Use the plots to quickly obtain the percent overshoot and the time of this event for the following second-order system:

$$P(s) = \frac{1}{s^2 + 2s + 4}$$

6.11. The second-order system in Problem 6.10 becomes the embedded second-order system in the third-order system given below.

  (a) If $R(s) = 1/s$, find and graph $y(t)$ for various values of a $(0 < a < 10)$. Indicate the percent overshoot and the time at which it occurs in each case. Verify that these quantities approach the values in Problem 6.10 if $a \geq 10$. In general this implies that $a \geq 10\zeta_n\omega_n$.

  (b) What do you think would happen if the transfer function had zeroes very close to the second-order poles?

$$P(s) = \frac{Y(s)}{R(s)} = \frac{1}{(s^2 + 2s + 4)(s + a)}$$

6.12. Figure P6.12 represents the pole-zero pattern for a control system for all values of gain K.

  (a) Find the minimum damping ratio for the system.

  (b) Find the system damped and undamped natural frequency at the calculated system damping ratio.

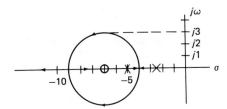

Figure P6.12   Pole-zero plot for Problem 6.12

Continuous

7

Continuous Proportional Integral Derivative (P.I.D.) Control

7.1 INTRODUCTION

This form of control is used extensively in industry and yet is understood, for the most part, in a way that is limited and superficial. To fully understand this type of control it is imperative that each individual component be considered separately. The proportional component responds to present error and provides a decrease in system rise time. The integral component responds to past error by accumulating the area beneath the error curve. The integration process provides zero steady-state error with a tendency to increase overshoot and settling time. The derivative component looks at future error by responding to rate of change of error. This tends to decrease overshoot and settling time but at the same time is very susceptible to noise in the system.

7.2 PROPORTIONAL CONTROL

A typical proportionally controlled second-order system is illustrated in Fig. 7.1.

This type of compensation simply multiplies the present error of the system

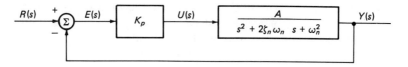

Figure 7.1 Proportionally controlled second-order system

with a constant value K_p. The system transfer function is given by

$$\frac{Y(s)}{R(s)} = \frac{A \cdot K_p}{s^2 + 2\zeta_n\omega_n s + (\omega_n^2 + A \cdot K_p)} \tag{7.2.1}$$

and the system characteristic equation is

$$s^2 + 2\zeta_n\omega_n s + (\omega_n^2 + A \cdot K_p) = 0 \tag{7.2.2}$$

The formal system characteristic equation from (6.2.5) is

$$s^2 + 2\zeta_m\omega_m s + \omega_m^2 = 0 \tag{7.2.3}$$

If we compare (7.2.2) with (7.2.3), we have the following:

$$\omega_m^2 = \omega_n^2 + A \cdot K_p \tag{7.2.4}$$

$$\zeta_m = \frac{\zeta_n\omega_n}{\omega_m} \tag{7.2.5}$$

From (7.2.4) and (7.2.5) the system undamped natural frequency (ω_m) increases while the system damping ratio (ζ_m) decreases with an increase in proportional gain, respectively. This implies that the system speed will increase with a tendency to overshoot.

System Error

The system error is given by

$$E(s) = R(s) - Y(s) \tag{7.2.6}$$

which simplifies to

$$E(s) = R(s)\left[\frac{s^2 + 2\zeta_n\omega_n s + \omega_n^2}{s^2 + 2\zeta_n\omega_n s + (\omega_n^2 + A \cdot K_p)}\right] \tag{7.2.7}$$

and the steady-state error $e(\infty)$ due to a step input ($R(s) = 1/s$) is attained by applying the final-value theorem:

$$e(\infty) = \lim_{s \to 0} s \cdot E(s) \tag{7.2.8}$$

$$= \frac{\omega_n^2}{\omega_n^2 + AK_p} \tag{7.2.9}$$

From (7.2.9) it is evident that the system has an error present at all times which decreases as proportional gain (K_p) increases. The tradeoff in proportional control implies more overshoot for a smaller error.

Time Domain

Figure 7.2 summarizes our findings in the time domain.

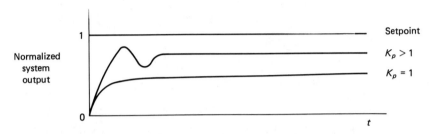

Figure 7.2   Time-domain response with increasing proportional gain

Frequency Domain

Also of interest is the migration of the system poles as seen in Fig. 7.3.

Analog Circuit

Proportional gain can be implemented with the operational amplifier configuration as seen in Fig. 7.4.
The proportional gain is given by

$$K_p = -\frac{R_2}{R_1} \qquad (7.2.10)$$

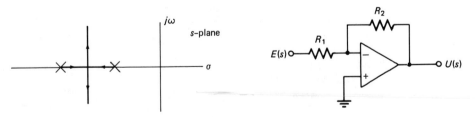

Figure 7.3   Migration of system poles with increasing proportional gain

Figure 7.4   Proportional gain circuit

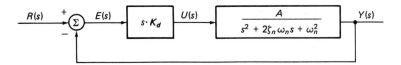

Figure 7.5 Derivative control of second-order system

7.3 DERIVATIVE CONTROL

A typical second-order system with derivative compensation is given in Fig. 7.5.

In this case the control scheme concentrates on the rate of change of error multiplied by a gain factor K_d which implies a future sense of control. The system transfer function is given by

$$\frac{Y(s)}{R(s)} = \frac{A \cdot K_d s}{s^2 + (2\zeta_n\omega_n + A \cdot K_d)s + \omega_n^2} \tag{7.3.1}$$

and the system characteristic equation is

$$s^2 + (2\zeta_n\omega_n + A \cdot K_d)s + \omega_n^2 = 0 \tag{7.3.2}$$

If we compare (7.3.2) with (7.2.3), we find that

$$\omega_m = \omega_n \tag{7.3.3}$$

$$\zeta_m = \zeta_n + \frac{A}{2\omega_m} \cdot K_d \tag{7.3.4}$$

From (7.3.3) and (7.3.4) we can see that the system undamped natural frequency (ω_m) is the same as the plant undamped natural frequency (ω_n), while the system damping ratio (ζ_m) increases as derivative gain (K_d) increases, respectively. This implies that we have control of the system overshoot and ringing with derivative control. It should be noted that this type of control is very susceptible to noise and is never used in its true form.

System Error

Using (7.2.6), the system error is given by

$$E(s) = R(s)\left[\frac{s^2 + 2\zeta_n\omega_n s + \omega_n^2}{s^2 + (2\zeta_n\omega_n + AK_d)s + \omega_n^2}\right] \tag{7.3.5}$$

and the steady-state error due to a step input, using (7.2.8), is

$$e(\infty) = \frac{\omega_n^2}{\omega_n^2} = 1 \qquad (7.3.6)$$

From (7.3.6) it is evident that this type of control is never used alone, as it promotes 100% error at steady state. It is important to look at this type of control in solidarity to grasp its tendency to reduce system overshoot and ringing.

Time Domain

Figure 7.6 depicts a typical system response with derivative control only.

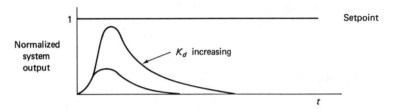

Figure 7.6 Time-domain response with increasing derivative gain

Frequency Domain

In Figure 7.7 we can see the system poles (assuming complex poles) move toward the real axis in a circular fashion. This allows the undamped natural frequency to remain constant while increasing the value of the system damping ratio (ζ_m).

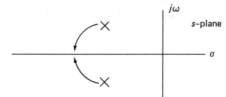

Figure 7.7 Migration of system poles with increasing derivative gain

Analog Circuit

Derivative gain can be implemented with the operational amplifier configuration as seen in Fig. 7.8.

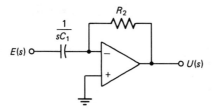

Figure 7.8 Derivative gain circuit

The derivative gain is given by

$$K_d = -C_1 R_2 \qquad (7.3.7)$$

7.4 INTEGRAL CONTROL

A typical second-order system with integral compensation is given in Fig. 7.9.

This type of compensation accumulates the area under the error curve and is therefore responsible for past type control. It is also multiplied by a constant gain K_i. The system transfer function is given by

$$\frac{Y(s)}{R(s)} = \frac{A \cdot K_i}{s^3 + 2\zeta_n \omega_n s^2 + \omega_n^2 s + A \cdot K_i} \qquad (7.4.1)$$

and the system characteristic equation is

$$s^3 + 2\zeta_n \omega_n s^2 + \omega_n^2 s + A \cdot K_i = 0 \qquad (7.4.2)$$

From (7.4.2) it is evident that our system has increased in order by one. A direct comparison to our second-order system is not evident, and it is necessary to disassemble this system into a first- and a second-order system. Since the first-order system decays with time only, the domineering response will be exhibited by the second-order system. If we assume the first-order system is

$$(s + a) \qquad (7.4.3)$$

and dividing (7.4.3) into (7.4.2) and allowing for zero remainder, since $(s + a)$ is a factor, we obtain the disassembled characteristic equation

$$(s + a)\left[s^2 + (2\zeta_n \omega_n - a)s + \frac{A \cdot K_i}{a}\right] = 0 \qquad (7.4.4)$$

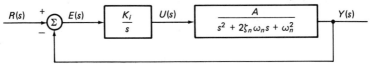

Figure 7.9 Integral control of second-order system

If we complete the multiplication of (7.4.4), we obtain

$$s^3 + 2\zeta_n\omega_n s^2 + \left[\frac{A \cdot K_i}{a} + a(2\zeta_n\omega_n - a)\right]s + A \cdot K_i = 0 \qquad (7.4.5)$$

At this point (7.4.2) and (7.4.5) represent the same quantity, and if this is to be the case as K_i increases, the value of a must also increase. If we look at (7.4.4), we find that as the value of a increases, the embedded second-order system will exhibit a low damping ratio and eventually become unstable. We can conclude that integral control promotes overshoot and, as derivative control, cannot be used alone.

System Error

Using (7.2.6), the system error is given by

$$E(s) = R(s)\left[\frac{s(s^2 + 2\zeta_n\omega_n s + \omega_n^2)}{s^3 + 2\zeta_n\omega_n s^2 + \omega_n^2 s + A \cdot K_i}\right] \qquad (7.4.6)$$

and the steady-state error due to a step input, using (7.2.8), is

$$e(\infty) = 0 \qquad (7.4.7)$$

From (7.4.7) it is obvious why integral control is included in systems. Aside from promoting overshoot and ringing, integral control is used to reduce the system error to zero.

Time Domain

Figure 7.10 provides a typical system response with integral control only.

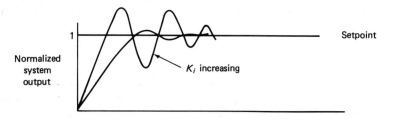

Figure 7.10 Time-domain response with increasing integral gain

Frequency Domain

In Figure 7.11 we can see the migration of the three system poles as the integral gain (K_i) increases. The pole on the real axis represents the disassembled first-order system, while the other two represent the second-order system. As the gain K_i increases, we can see that the first-order system decays to zero more rapidly, while the second-order system exhibits a lower damping ratio and follows a path to the right half of the s-plane. This, of course, leads to instability.

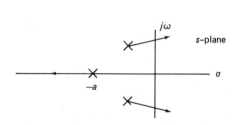

Figure 7.11 Migration of system poles with increasing integral gain

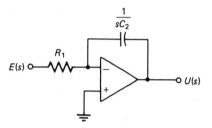

Figure 7.12 Integral gain circuit

Analog Circuit

Integral gain can be implemented with the operational amplifier configuration as seen in Fig. 7.12.

The integral gain is given by

$$K_i = -\frac{1}{R_1 C_2} \tag{7.4.8}$$

7.5 PROPORTIONAL DERIVATIVE CONTROL

A typical second-order system with proportional derivative compensation is given in Fig. 7.13.

This type of compensation provides both present and future control. As this type of control is of practical use, we will spend more time on its impli-

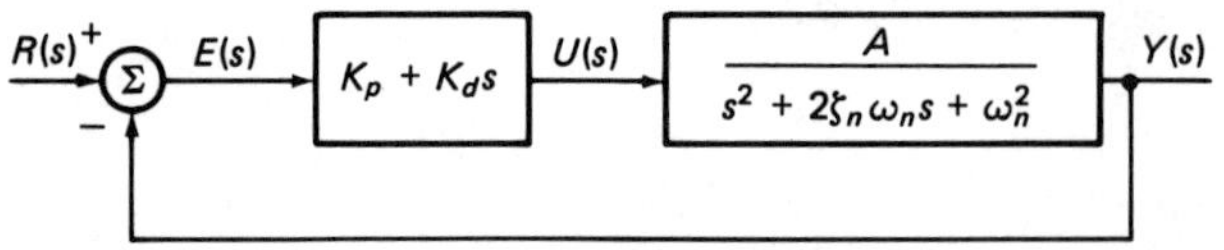

R(s) + Σ → E(s) → $K_p + K_d s$ → U(s) → $\dfrac{A}{s^2 + 2\zeta_n \omega_n s + \omega_n^2}$ → Y(s)

Figure 7.13 Proportional derivative control of second-order system

cations. The system transfer function is given by

$$\frac{Y(s)}{R(s)} = \frac{AK_d s + AK_p}{s^2 + (2\zeta_n \omega_n + AK_d)s + (\omega_n^2 + A \cdot K_p)} \tag{7.5.1}$$

and the system characteristic equation is

$$s^2 + (2\zeta_n \omega_n + AK_d)s + (\omega_n^2 + A \cdot K_p) = 0 \tag{7.5.2}$$

If we compare (7.5.2) with (7.2.3), we have

$$\omega_m^2 = \omega_n^2 + A \cdot K_p \tag{7.5.3}$$

$$\zeta_m = \frac{2\zeta_n \omega_n + AK_d}{2 \sqrt{\omega_n^2 + A \cdot K_p}} \tag{7.5.4}$$

From (7.5.3) we find that the system undamped natural frequency (ω_m) increases as proportional gain (K_p) increases. This provides an avenue for system speed increase. From (7.5.4) it is evident that proportional gain also tends to decrease the system damping ratio (ζ_m) by approximately the inverse of its square root. At the same time the system damping ratio can be increased directly by increasing the derivative gain (K_d). This implies that we can tune this system and that proportional gain can be changed substantially while derivative gain only moderately. If we are to tune this system properly, we must look closer at the implications of the system step response. If we assume that our system has enough proportional gain such that $0 < \zeta_m < 1$, and using (7.5.1) with inverse Laplace transforms, the following represents a typical step response in time ($y(t)$):

$$y(t) = K_1 + e^{-\left(\frac{2\zeta_n \omega_n + AK_d}{2}\right)t}[K_2 \cos K_3 t + K_4 \sin K_3 t] \tag{7.5.5}$$

where K_1, K_2, K_3, K_4 are the appropriate constants.
Studying (7.5.5), we find that our system settling time T_s can be modified by the damped exponential. In fact we can modify the time constant (τ) which is

$$\tau = \frac{2}{2\zeta_n \omega_n + A \cdot K_d} \tag{7.5.6}$$

and, defining our system settling time as five time constants, we have

$$T_s = 5\tau = \frac{10}{2\zeta_n \omega_n + A \cdot K_d} \tag{7.5.7}$$

From (7.5.7) we can calculate the derivative gain, given the system settling time. The system damping ratio can also be specified and, using (7.5.4), the proportional gain can be calculated. This provides a mathematical method of tuning this system so that it will meet the response specifications.

System Error

Using (7.2.6), the system error is given by

$$E(s) = R(s) \left[\frac{s^2 + 2\zeta_n \omega_n s + \omega_n^2}{s^2 + (2\zeta_n \omega_n + A \cdot K_d)s + (\omega_n^2 + A \cdot K_p)} \right] \qquad (7.5.8)$$

and the steady-state error due to a step input, using (7.2.8), is

$$e(\infty) = \frac{\omega_n^2}{\omega_n^2 + A \cdot K_p} \qquad (7.5.9)$$

The value of the error given in (7.5.9) is the same as in (7.2.9), which indicates a similar error as proportional gain only. If we look more closely at (7.5.4), it is evident that this type of control system will exhibit higher proportional gain and therefore have a smaller error as compared to proportional control only.

Time Domain

At this point it is much easier to understand the time response by actually going through a design example.

EXAMPLE 7.1

A robotics firm has purchased a DC motor with the following shaft speed ($\Omega(s)$) to armature voltage ($V_a(s)$) transfer function ($G(s)$) and has decided to use it in a constant-speed conveyor system.

$$G(s) = \frac{\Omega(s)}{V_a(s)} = \frac{2}{s^2 + 4s + 3}$$

It is a design requirement that the motor will exhibit a settling time of 1 second ($T_s = 1$ sec) and a damping ratio of 0.707 ($\zeta_m = 0.707$) as a result of a step input. For a first iteration it is decided that a proportional derivative control scheme will be used. Calculate the gains (K_p, K_d) required, the system error, and provide a graph of the system response due to a step setpoint change (normalized).

SOLUTION   It is good practice to provide a block diagram of the control system and list the information both given and required.   First the diagram and information given:

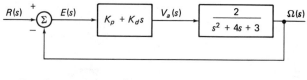

T_s = 1 sec	K_p = ?	$2\zeta_n\omega_n = 4$	A = 2
ζ_m = 0.707	K_d = ?	$\omega_n = \sqrt{3}$ rad/sec	

Using (7.5.1), the system transfer function is

$$\frac{\Omega(s)}{R(s)} = \frac{2K_d s + 2K_p}{s^2 + (4 + 2K_d)s + (3 + 2K_p)}$$

and the system characteristic equation is

$$s^2 + (4 + 2K_d)s + (3 + 2K_p) = 0$$

Since the settling time is 1 second, we can calculate the gain K_d:

$$1 = \frac{10}{4 + 2K_d}$$

$$\boxed{K_d = 3}$$

and, using the specified damping ratio ζ_m, we calculate K_p:

$$0.707 = \frac{4 + (2)(3)}{2\sqrt{3 + 2K_p}}$$

$$\boxed{K_p = 23.5}$$

The steady-state error is

$$\boxed{e(\infty) = \frac{3}{3 + 2(23.5)} = 0.06}$$

The system transfer function becomes

$$\frac{\Omega(s)}{R(s)} = \frac{6s + 47}{s^2 + 10s + 50}$$

and the step response ($R(s)$ = $1/s$) is

$$\omega(t) = \mathcal{L}^{-1}\left[\frac{6s + 47}{s(s^2 + 10s + 50)}\right]$$

$$\omega(t) = 0.94 - 0.94e^{-5t}\cos 5t + 0.26e^{-5t}\sin 5t$$

Figure 7.14 provides a graph for the system response. From this graph it is evident that our specifications have been realized.

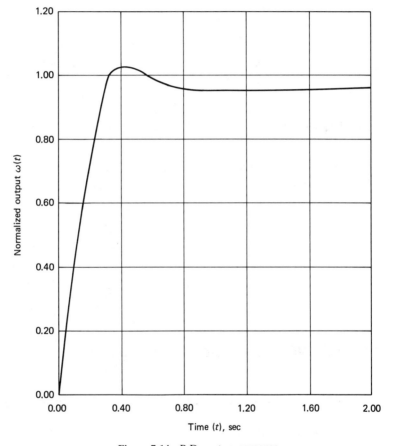

Figure 7.14   P.D. system response

Frequency Domain

Figure 7.15 follows the path of the system poles with both gains at zero and letting the proportional gain attain its final value. Then the derivative gain is applied until it reaches its final value. This allows the reader to see the con-

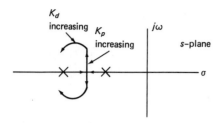

Figure 7.15 Migration of system poles with increasing proportional derivative gain

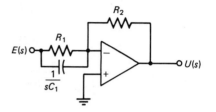

Figure 7.16 Proportional derivative gain circuit

tribution of each type of control as viewed in the frequency domain. As expected, the individual contributions are a combination of Fig. 7.3 and Fig. 7.7.

Analog Circuit

The operational amplifier circuit in Fig. 7.16 can be used to implement proportional derivative gain.

The proportional gain is given by

$$K_p = -\frac{R_2}{R_1} \tag{7.5.10}$$

with the derivative gain being

$$K_d = -C_1 R_2 \tag{7.5.11}$$

7.6 PROPORTIONAL INTEGRAL CONTROL

A typical second-order system with proportional integral compensation is shown in Fig. 7.17

This type of compensation provides both past and present control. Since the past sense of control is present, this system will require an optimization scheme which will allow the quickest response with minimum overshoot and ringing. The system transfer function is given by

$$\frac{Y(s)}{R(s)} = \frac{A \cdot K_p s + A \cdot K_i}{s^3 + 2\zeta_n \omega_n s^2 + (\omega_n^2 + A \cdot K_p)s + A \cdot K_i} \tag{7.6.1}$$

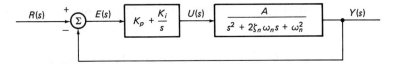

Figure 7.17   Proportional integral control of a second-order system

and the system characteristic equation is

$$s^3 + 2\zeta_n\omega_n s^2 + (\omega_n^2 + A \cdot K_p)s + A \cdot K_i = 0 \qquad (7.6.2)$$

From (7.6.2) we find that our system contains a first- and a second-order system. If the first-order system is taken as before $(s + a)$, then by long division the second-order system will appear.   The form of the second-order system is similar to that found in equation (7.4.4).   If the damping ratio of the second-order system is between zero and one, then the optimal location of the first-order pole is equivalent to the real part of the pole locations for the second-order system.   This will guarantee the quickest response possible with this type of control scheme.   Mathematically this is equivalent to

$$-a = -\frac{2\zeta_n\omega_n - a}{2} \qquad (7.6.3)$$

which reduces to

$$a = \frac{2\zeta_n\omega_n}{3} \qquad (7.6.4)$$

For an optimal system the first-order system should be

$$\left(s + \frac{2\zeta_n\omega_n}{3}\right) \qquad (7.6.5)$$

and by using long division the second-order system will emerge in terms of the gains K_p and K_i.   From the specified system damping ratio (ζ_m) and by setting the remainder to zero [since (7.6.5) is a factor] we can calculate our optimal values for K_p and K_i.   An example given below will provide a better understanding of this process.

System Error

Using (7.2.6), the system error is given by

$$E(s) = R(s)\left[\frac{s(s^2 + 2\zeta_n\omega_n s + \omega_n^2)}{s^3 + 2\zeta_n\omega_n s^2 + (\omega_n^2 + A \cdot K_p)s + A \cdot K_i}\right] \qquad (7.6.6)$$

and the steady-state error due to a step input, using (7.2.8), is

$$e(\infty) = 0 \tag{7.6.7}$$

As expected from (7.6.7), the system error will be zero. The integration process remains in effect until the error diminishes to zero. The only tradeoff in this type of control is speed of response, as we will find out in the next example.

Time Domain

In order to fully understand this type of control we will repeat Example 7.1, and, since we are designing an optimal control, it will be interesting to see how well it compares to proportional derivative control.

EXAMPLE 7.2

The block diagram below depicts the control system.

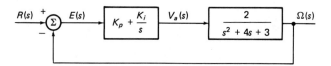

Using (7.6.1), the system transfer function is

$$\frac{\Omega(s)}{R(s)} = \frac{2K_p s + 2K_i}{s^3 + 4s^2 + (3 + 2K_p)s + 2K_i}$$

and the system characteristic equation is

$$s^3 + 4s^2 + (3 + 2K_p)s + 2K_i = 0$$

For an optimal system the first-order system should be

$$s + 1.333$$

and, using long division,

$$
\begin{array}{r}
s^2 + 2.667s + (2K_p - 0.555) \\
s + 1.333 \overline{\smash{\big)}\ s^3 + \quad 4s^2 \quad + (3 + 2K_p)s + 2K_i} \\
\underline{s^3 + 1.333s^2} \\
2.667s^2 \quad + (3 + 2K_p)s \\
\underline{2.667s^2 \quad + \quad\quad 3.555s} \\
(2K_p - 0.555)s + 2K_i \\
\underline{(2K_p - 0.555)s + (2.666K_p - 0.740)} \\
2K_i - 2.666K_p + 0.740
\end{array}
$$

the second-order system is

$$s^2 + 2.667s + (2K_p - 0.555) \qquad (1)$$

with the remainder equal to zero [since $(s + 1.333)$ is a factor]:

$$2K_i - 2.666K_p + 0.740 = 0 \qquad (2)$$

From (1) we have

$$2\zeta_m\omega_m = 2.667$$

$$\omega_m = 1.886 \qquad \text{rad/sec} \qquad (\zeta_m = 0.707)$$

$$\omega_m^2 = 3.558$$

$$2K_p - 0.555 = 3.558$$

$$\boxed{K_p = 2.056}$$

Substitute into (2):

$$\boxed{K_i = 2.371}$$

The steady-state error will be zero due to integral content. The system transfer function becomes

$$\frac{\Omega(s)}{R(s)} = \frac{4.112s + 4.742}{s^3 + 4s^2 + 7.112s + 4.742}$$

and the step response $(R(s) = 1/s)$ is

$$\omega(t) = \mathcal{L}^{-1}\left[\frac{4.112s + 4.742}{s(s^3 + 4s^2 + 7.112s + 4.742)}\right]$$

$$\boxed{\omega(t) = 1 + 0.312e^{-1.333t} - 1.312e^{-1.33t} \cos 1.333t - e^{-1.333t} \sin 1.333t}$$

Figure 7.18 provides a graph of both the proportional derivative and proportional integral system responses.

From Figure 7.18 it is evident that our optimized proportional integral control scheme is much slower as compared to the proportional derivative control. The loss of speed results in zero error control, which in itself is very important. If the system is to be any faster, then it will have to tolerate more overshoot and ringing. Also of interest are the magnitudes of both gains, in that they are very close to each other. As we continue with our control schemes, we will find a general tracking of these two gains in terms of order of magnitude. Readers are invited to see if they can generate a quicker proportional integral control scheme satisfying the specifications of this particular problem.

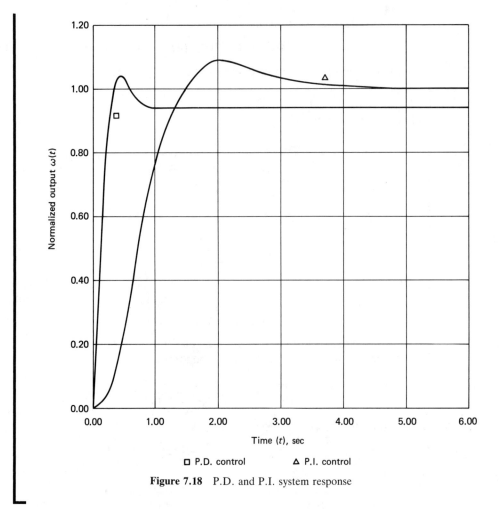

Figure 7.18 P.D. and P.I. system response

Frequency Domain

Figure 7.19 follows the path of the system poles with both gains at zero and letting the integral gain attain its final value.   Then the proportional gain is increased to its final value.   It is evident that the integral gain moves the system toward instability, while the proportional gain pulls the system back into the left half of the s-plane.   This helps us understand why integral control cannot be used alone.

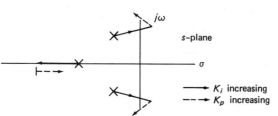
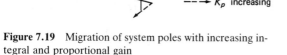

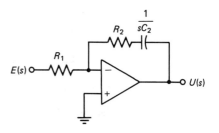

Figure 7.19   Migration of system poles with increasing integral and proportional gain

Figure 7.20   Proportional integral gain circuit

Analog Circuit

The operational amplifier circuit in Fig. 7.20 can be used to implement proportional integral gain.

The proportional gain is given by

$$K_p = -\frac{R_2}{R_1} \qquad (7.6.8)$$

with the integral gain being

$$K_i = -\frac{1}{R_1 C_2} \qquad (7.6.9)$$

7.7 PROPORTIONAL INTEGRAL DERIVATIVE CONTROL

A typical second-order system with proportional integral derivative compensation is shown in Fig. 7.21.

This type of compensation provides past, present, and future control. From the individual components presented previously this type of control should provide zero error, fast response, and controllable ringing due to integral, proportional and derivative gain, respectively.   The system transfer function is given by

$$\frac{Y(s)}{R(s)} = \frac{s^2 A \cdot K_d + s A \cdot K_p + A \cdot K_i}{s^3 + s^2 (2\zeta_n \omega_n + A \cdot K_d) + s(\omega_n^2 + A \cdot K_p) + A \cdot K_i} \qquad (7.7.1)$$

and the system characteristic equation is

$$s^3 + s^2 (2\zeta_n \omega_n + A \cdot K_d) + s(\omega_n^2 + A \cdot K_p) + A \cdot K_i = 0 \qquad (7.7.2)$$

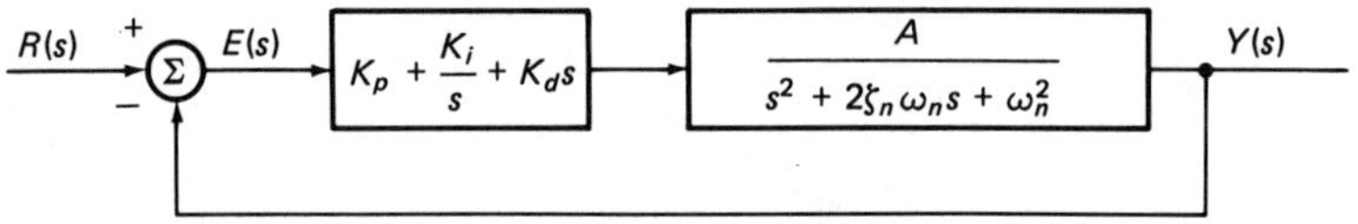

Figure 7.21 Proportional integral derivative control of second-order system

If we disassemble (7.7.2) into its first- and second-order systems, we obtain

$$(s + a)\left[s^2 + s(2\zeta_n\omega_n + A \cdot K_d - a) + \frac{A \cdot K_i}{a}\right] = 0 \qquad (7.7.3)$$

with the second-order system being

$$s^2 + s(2\zeta_n\omega_n + A \cdot K_d - a) + \frac{A \cdot K_i}{a} \qquad (7.7.4)$$

Now, referring to Section 7.6, the location of the first-order system should be

$$-a = -\frac{2\zeta_n\omega_n + A \cdot K_d - a}{2} \qquad (7.7.5)$$

or

$$a = \frac{2\zeta_n\omega_n + A \cdot K_d}{3} \qquad (7.7.6)$$

to provide optimum response due to integral content. From Section 7.5 we found that the derivative content provided control over system settling time T_s. Applying this idea to this system, we have

$$T_s = \frac{10}{2\zeta_n\omega_n + AK_d - a} \qquad (7.7.7)$$

Using (7.7.6) and (7.7.7), we can solve for the values of a and K_d. As in proportional integral control we now use long division to calculate the gains K_p and K_i. This implies that we can design a system with zero error and still have control over system settling time. An example given below will help the reader appreciate the fine details just covered.

System Error

Using (7.2.6), the system error is given by

$$E(s) = R(s)\left[\frac{s^3 + 2\zeta_n\omega_n s^2 + \omega_n^2 s}{s^3 + (2\zeta_n\omega_n + AK_d)s^2 + (\omega_n^2 + A \cdot K_p)s + AK_i}\right] \qquad (7.7.8)$$

and the steady-state error due to a step input, using (7.2.8), is

$$e(\infty) = 0 \qquad (7.7.9)$$

Time Domain

EXAMPLE 7.3

Let's design a P.I.D. control system for Example 7.1. The block diagram below depicts the control scheme.

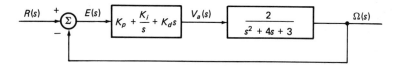

Using (7.7.1), the system transfer function is

$$\frac{\Omega(s)}{R(s)} = \frac{s^2 \cdot 2K_d + s \cdot 2K_p + 2K_i}{s^3 + s^2(4 + 2K_d) + s(3 + 2K_p) + 2K_i}$$

and the system characteristic equation is

$$s^3 + s^2(4 + 2K_d) + s(3 + 2K_p) + 2K_i = 0$$

For an optimal system the first-order system should be

$$a = \frac{4 + 2K_d}{3} \qquad (1)$$

and to insure a settling time of 1 second we have

$$1 = \frac{10}{4 + 2K_d - a} \qquad (2)$$

Using equations (1) and (2), we have

$$\boxed{a = 5}$$

$$\boxed{K_d = 5.5}$$

Now, using long division,

$$
\begin{array}{r}
s^2 \quad + 10s + (2K_p - 47) \\
s + 5 \overline{\smash{\big)}\, s^3 + 15s^2 \quad + s(3 + 2K_p) + 2K_i} \\
\underline{s^3 + 5s^2} \\
10s^2 \quad + s(3 + 2K_p) \\
\underline{10s^2 \quad + s(50)} \\
s(2K_p - 47) + 2K_i \\
\underline{s(2K_p - 47) + 10K_p - 235} \\
2K_i - 10K_p + 235
\end{array}
$$

The second-order system is

$$s^2 + 10s + (2K_p - 47) \qquad (3)$$

with the remainder equal to zero:

$$2K_i - 10K_p + 235 = 0 \qquad (4)$$

From (3) we have

$$2\zeta_m\omega_m = 10$$

$$\omega_m = 7.07 \quad \text{rad/sec} \quad (\zeta_m = 0.707)$$

$$\omega_m^2 = 50$$

$$2K_p - 47 = 50$$

$$\boxed{K_p = 48.5}$$

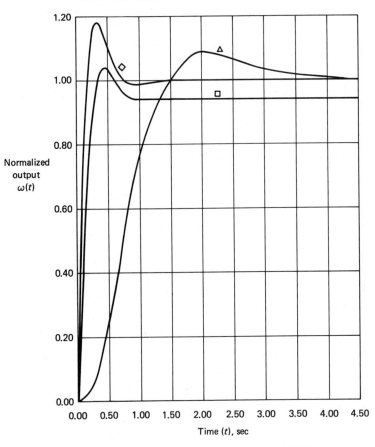

□ P.D. control   △ P.I. control   ◇ P.I.D. control

Figure 7.22   P.D., P.I., and P.I.D. system response

Substitute into (4), and

$$2K_i - 10(48.5) + 235 = 0$$

$$\boxed{K_i = 125}$$

The steady-state error is zero due to integral content. The system transfer function becomes

$$\frac{\Omega(s)}{R(s)} = \frac{11s^2 + 97s + 250}{s^3 + 15s^2 + 100s + 250}$$

and the step response $(R(s) = 1/s)$ is

$$\omega(t) = \mathcal{L}^{-1}\left[\frac{11s^2 + 97s + 250}{s(s^3 + 15s^2 + 100s + 250)}\right]$$

$$\boxed{\omega(t) = 1 - 0.32e^{-5t} - 0.68e^{-5t}\cos 5t + 1.2e^{-5t}\sin 5t}$$

Figure 7.22 provides a graph of all three methods of control for the system in Example 7.1. From this graph it is evident that proportional integral derivative type control provides an exceptional response with zero error. The reader is asked to study Fig. 7.23 carefully and try to predict the system response if any one of the gains were to be changed by maintenance personnel.

Frequency Domain

Figure 7.23 follows the path of the system poles with all three gains starting at zero and each individually reaching its final value.

From Fig. 7.22 we can see why a system with P.I.D. control requires tuning. Each gain provides a specific path for the system poles to follow in the s-plane, and it is imperative that the control technologist understand the implications in both the frequency and the time domain. We will design another P.I.D. control system shortly and this time investigate the effects on system response of increasing any one gain at a time. You have been asked

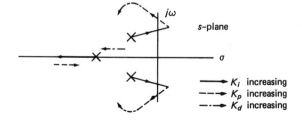

K_i increasing
K_p increasing
K_d increasing

Figure 7.23 Migrating system poles with increasing integral proportional system

to predict the results already and at this point should be quite confident with your answer.

Analog Circuits

The operational amplifier circuits in Fig. 7.24 can be used to implement proportional integral derivative gain.

The proportional gain is given by

$$\text{(a)} \quad K_p = -\left(\frac{R_2}{R_1} + \frac{C_1}{C_2}\right) \tag{7.7.10}$$

$$\text{(b)} \quad K_p = -\left(\frac{R_2}{R_1} + \frac{R_4}{R_3}\right) \tag{7.7.11}$$

The integral gain is given by

$$\text{(a)} \quad K_i = -\frac{1}{R_1 C_2} \tag{7.7.12}$$

$$\text{(b)} \quad K_i = \frac{1}{R_1 C_2} \tag{7.7.13}$$

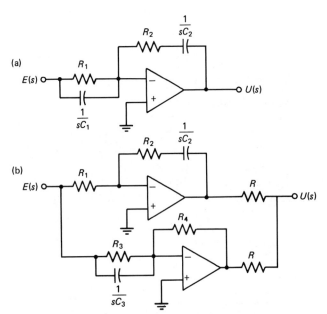

Figure 7.24 Proportional integral derivative gain circuits

The derivative gain is given by

$$\text{(a)} \qquad K_d = -C_1R_2 \qquad\qquad (7.7.14)$$

$$\text{(b)} \qquad K_d = -C_3R_4 \qquad\qquad (7.7.15)$$

The circuit in Fig. 7.24(b) is included to provide more flexibility in choosing the individual gains. The value of resistor R depends on the external circuitry and is included only to show the summing aspect of the individual gains.

EXAMPLE 7.4

We will now examine a realistic problem of level control often found in the chemical processing industry. The setup is shown below and consists of a two-tank level system. Tank 1 receives liquid via a control valve and passes the liquid to Tank 2 through a connecting pipe, and the liquid is drawn out of Tank 2 through a similar pipe. A level transducer is located in Tank 2 which provides the required feedback to control the level in this tank. This setup allows drastic changes in Tank 1 while transmitting minimum disturbances to Tank 2.

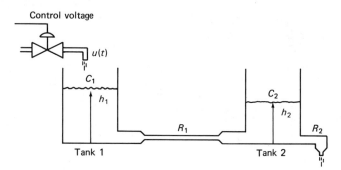

Sometimes it is beneficial to represent a physical system with an equivalent electrical circuit. Each tank can be represented as a capacitor and each pipe as a resistance. We are actually changing the incoming current to maintain a constant voltage on the second capacitor, and our transfer function for the plant becomes the ratio of this voltage to input current ratio (see the next diagram).

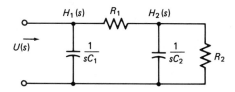

Using simple circuit analysis, we obtain

$$H_1(s)C_1s = \frac{H_2(s) - H_1(s)}{R_1} + U(s) \tag{1}$$

$$H_2(s)C_2s = -\frac{H_2(s) - H_1(s)}{R_1} - \frac{H_2(s)}{R_2} \tag{2}$$

where the variables in Laplace are

C_1 = capacitance of Tank 1
C_2 = capacitance of Tank 2
$H_1(s)$ = height of liquid in Tank 1
$H_2(s)$ = height of liquid in Tank 2
R_1, R_2 = resistance to flow in pipes
$U(s)$ = input flow, and we will assume this to be the same as the voltage on the valve (control action) (i.e., 0 V provides no flow and 15 V is maximum flow)

— all with appropriate units.

We would like to control the level of Tank 2 using a P.I.D. control scheme. If the tank is subjected to a unit step input ($R(s) = 1/s$), then the level in the tank should exhibit a damping ratio of 0.707 and settle in 2 seconds. A graph of the response and the control action is required. The gains should be calculated so as to promote an optimal response. The block diagram for the control scheme is given below.

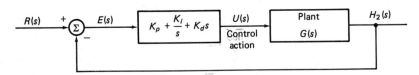

Using (1) and (2) with $C_1 = C_2 = 1$, $R_1 = 0.5$, $R_2 = 0.333$, we have

$$sH_1(s) = 2(H_2(s) - H_1(s)) + U(s) \tag{3}$$

$$sH_2(s) = -2(H_2(s) - H_1(s)) - 3H_2(s) \tag{4}$$

Solve for $H_1(s)$ in (4), substitute into (3), and simplify to obtain the plant transfer function $G(s)$:

$$G(s) = \frac{H_2(s)}{U(s)} = \frac{2}{s^2 + 7s + 6}$$

The system transfer function is

$$\frac{H_2(s)}{R(s)} = \frac{2K_d s^2 + 2K_p s + 2K_i}{s^3 + s^2(7 + 2K_d) + s(6 + 2K_p) + 2K_i}$$

with the characteristic equation being

$$s^3 + s^2(7 + 2K_d) + s(6 + 2K_p) + 2K_i = 0$$

For optimal response the first-order system is given by

$$a = \frac{7 + 2K_d}{3} \tag{5}$$

and to insure a 2-second settling time we have

$$2 = \frac{10}{7 + 2K_d - a} \tag{6}$$

Using equations (5) and (6), we have

$$\boxed{a = 2.5}$$

$$\boxed{K_d = 0.25}$$

and, performing long division,

$$
\begin{array}{r}
s^2 + 5s + (2K_p - 6.5) \\
s + 2.5 \overline{\smash{\big)}\ s^3 + s^2(7.5) \quad + s(6 + 2K_p) + 2K_i} \\
\underline{s^3 + s^2(2.5)} \\
s^2(5) \quad + s(6 + 2K_p) \\
\underline{s^2(5) \quad + s(12.5)} \\
s(2K_p - 6.5) + 2K_i \\
\underline{s(2K_p - 6.5) + 5K_p - 16.25} \\
2K_i - 5K_p + 16.25
\end{array}
$$

The second-order system is

$$s^2 + 5s + (2K_p - 6.5) \tag{7}$$

with the remainder equal to zero:

$$2K_i - 5K_p + 16.25 = 0 \tag{8}$$

From (7) we have

$$2\zeta_m \omega_m = 5$$

$$\omega_m = 3.54 \qquad \text{rad/sec} \qquad (\zeta_m = 0.707)$$

$$\omega_m^2 = 12.5$$

$$2K_p - 6.5 = 12.5$$

$$\boxed{K_p = 9.5}$$

Substitute into (8) and obtain

$$2K_i - 5(9.5) + 16.25 = 0$$

$$\boxed{K_i = 15.63}$$

The system transfer function becomes

$$\frac{H_2(s)}{R(s)} = \frac{0.5s^2 + 19s + 31.25}{s^3 + 7.5s^2 + 25s + 31.25}$$

and the step response $(R(s) = 1/s)$ is

$$h_2(t) = \mathcal{L}^{-1}\left[\frac{0.5s^2 + 19s + 31.25}{s(s^3 + 7.5s^2 + 25s + 31.25)}\right]$$

$$\boxed{h_2(t) = 1 + 0.84e^{-2.5t} - 1.84e^{-2.5t}\cos 2.5t - 0.8e^{-2.5t}\sin 2.5t}$$

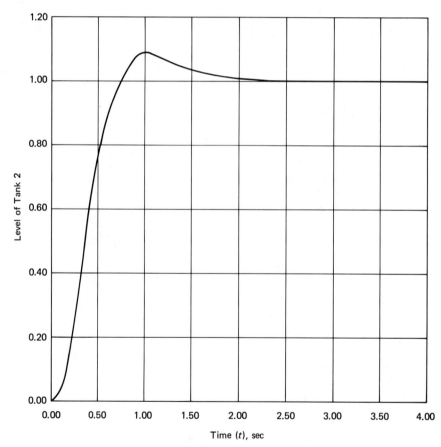

Figure 7.25 Graph of system response

Figure 7.25 provides a graph for the system response, and it is evident that our system has met the specifications in terms of damping ratio and settling time. It is left as an exercise for the reader to calculate and plot the level in Tank 1, given the same specifications and step input.

To calculate the control action we need to go back to our initial block diagram. From this we have

$$U(s) = E(s) \left[K_p + \frac{K_i}{s} + K_d s \right]$$

$$= (R(s) - Y(s)) \left[K_p + \frac{K_i}{s} + K_d s \right]$$

Substituting all the known values and simplifying ($t > 0$),

$$U(s) = \frac{9.375s^3 + 77.375s^2 + 158.56s + 93.75}{s(s^3 + 7.5s^2 + 25s + 31.25)}$$

Taking the inverse Laplace transform, we have

$$u(t) = 3 - 2.21e^{-2.5t} + 8.58e^{-2.5t} \cos 2.5t + 9.2e^{-2.5t} \sin 2.5t$$

Figure 7.26 provides a graph for the control action, and it is evident that the

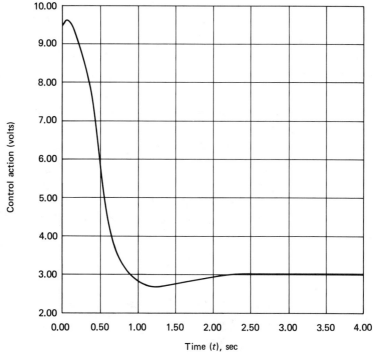

Figure 7.26   Graph of system control action (voltage on control valve)

valve immediately opens to approximately 63% and in 1 second dips to 17% (based on full opening at 15 V). From Fig. 7.25 we can see that at this time the level in the tank exceeds its setpoint. The level in the tank drops to its setpoint and the valve increases and stabilizes at 20% opening. The system settles in 2 seconds, as required in the specifications.

Using our optimal system, we will now change the individual gains, one at a time, leaving the others constant, and graph the corresponding responses. For ease of calculation we will simply double the gain in question.

Double K_p

In this case K_d and K_i are as calculated in our optimal P.I.D. system. The transfer function becomes

$$\frac{H_2(s)}{R(s)} = \frac{0.5s^2 + 38s + 31.25}{s^3 + 7.5s^2 + 44s + 31.25}$$

and, taking the inverse Laplace transform, the time step response is

$$h_2(t) = 1 - 0.02916e^{-0.8099t} - 0.971e^{-3.345t} \cos 5.2337t$$

$$- 0.5295e^{-3.345t} \sin 5.2337t$$

Double K_d

In this case K_p and K_i are as calculated in our optimal P.I.D. system. The transfer function becomes

$$\frac{H_2(s)}{R(s)} = \frac{s^2 + 19s + 31.25}{s^3 + 8s^2 + 25s + 31.25}$$

and, taking the inverse Laplace transform, the time step response is

$$h_2(t) = 1 + 1.288e^{-3.286t} - 2.288e^{-2.357t} \cos 1.99t - 0.0806e^{-2.357t} \sin 1.99t$$

Double K_i

In this case K_p and K_d are as calculated in our optimal P.I.D. system. The transfer function becomes

$$\frac{H_2(s)}{R(s)} = \frac{0.5s^2 + 19s + 62.5}{s^3 + 7.5s^2 + 25s + 62.5}$$

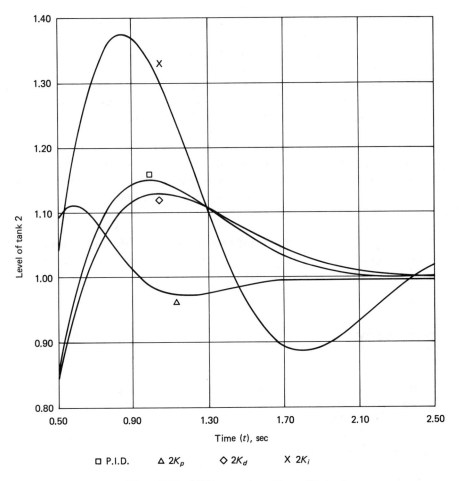

Figure 7.27   P.I.D. response with modified gains

and, taking the inverse Laplace transform, the time step response is

$$h_2(t) = 1 + 0.16e^{-5t} - 1.16e^{-1.25t} \cos 3.307t - 0.0454e^{-1.25t} \sin 3.307t$$

Figure 7.27 provides an expanded graph of all the modified and original P.I.D. responses. It is evident that the only response that meets the specifications is the original optimal system. To explain the difference in responses, which you should have predicted, we have to go back to Fig. 7.22 and study the migration of the poles in the system.

If K_p is increased, we find the pole on the real axis moving toward the origin. The other two poles move to the left and away from the horizontal axis. From the pole on the real axis it is evident that the system exhibits a longer settling time, and from the other two poles the embedded second-order system is faster with a lower damping ratio.   The result is a fast response with overshoot, undershoot, and a long settling time.

If K_i is increased, we find the pole on the real axis migrating to the left. The other two poles move toward the vertical axis and away from the horizontal axis.   The pole on the real axis decays quickly, and the response is characteristic of the embedded second-order system.   From the second-order system we can expect a high-speed system with very low damping ratio.

This case is the most interesting and often is misunderstood.   Did you get it right?   If K_d is increased, one would expect a quicker system, and if our system had not been optimized, this would be the case.   The pole on the real axis migrates to the left and decays very quickly, and again the system response is characteristic of the embedded second-order system.   The poles of the second-order system move in a spiral path toward the origin, which implies a sluggish system.

Again it is very important that the reader fully understand the frequency domain and the implications of the individual gains.   If this is not the case, it is requested the reader revisit this chapter.

7.8 SUMMARY

In this chapter we investigated proportional, integral, and derivative control as separate entities.   In proportional control we found a constant error present at all times, which diminished as gain increased.   In integral control we found a tendency toward instability as gain increased, yet it promoted zero error control.   In derivative control we found that the system speed could be retained with the possibility of decreasing system damping ratio.   It was evident that neither integral nor derivative control could be used alone.   As we combined proportional and derivative control, it was evident that our system was quick in response with control over damping ratio.   The proportional gain required in this system was much higher than that found in proportional control only.   Since the error had the same expression as that found in proportional control only, we found a smaller steady-state error.   The combination of proportional plus integral control provided zero-error control.   The price for this property was reflected in system speed.   To increase the speed, an optimal design scheme was initiated.   The final incorporation of proportional plus integral plus derivative control provided an exceptional control system with zero error and complete control over system settling time.   This system also required an optimal design scheme.   As each gain was changed individually, we found the corresponding change in system response.   The only surprise came with an increase of derivative gain only, which provided a slower system.   This was a result of an optimal system design and the spiral path of the system poles.

In all modes of control we tracked the migration of the system poles with an emphasis on the response as seen from the frequency domain. The time-domain responses were also presented. In order to implement the various modes of control, the equivalent operational amplifier circuits were provided with the necessary gain equations.

Problems

7.1. What is the major drawback with proportional control?

7.2. The implementation of the true derivative gain circuit as given in Fig. 7.8 is impractical. If we find the transfer function for this circuit, we have

$$\frac{U(s)}{E(s)} = sR_2C_1$$

Generate and plot the frequency response for this transfer function ($R_2 = 1\ M\Omega$, $C_1 = 1\ \mu F$). Why is this circuit impractical? Consider the practical derivative gain circuit given in Fig. P7.2. Generate the transfer function for this circuit. Plot the frequency response ($R_1 = 1\ M\Omega$) with all other values as previously used. Why is this circuit better?

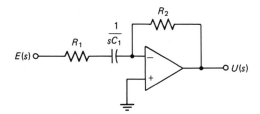

Figure P7.2 Practical derivative circuit for Problem 7.2

7.3. Integral control promotes zero error. This implies that the input to the circuit in Fig. 7.12 is zero. If the process is to continue, it requires a nonzero control action. Provide an explanation as to how this circuit can provide a finite output with a zero input.

7.4. A system is operating with proportional control only. The steady-state error is measured at 20% with $K_p = 15$. The system is modified to contain derivative control also. If $K_p = 15$, $K_d = 2$, what is the new steady-state error? What are the benefits of adding the derivative control?

7.5. If the circuit in Fig. 7.16 is to be practical, how should it be modified? Why?

7.6. If the circuits in Fig. 7.24 are to be practical, how should they be modified?

7.7. In Problem 6.11 we found an approximation for the location of the first-order pole with respect to the locations of the second-order poles if we are to consider the second-order system dominant. Why is this not the case with P.I.D. control? The answer lies in the migration of the poles relative to each gain. What about the system zeroes?

7.8. A P.I.D. system is operating normally and someone tampers with the gains. After a step change in setpoint, the system is found to have large overshoot and sustained oscillations. What are the possibilities in the tamperings?

7.9. What is the effect on bandwidth if the integral gain is increased in a P.I.D. control system?

7.10. What is the effect on system damping ratio if the proportional gain is increased in a P.I.D. control system?

7.11. What is the effect on rise time if the derivative gain is decreased in a P.I.D. control system?

7.12. The following plant is proportionally controlled. Find the proportional gain required if the steady-state error is to be 6%. What are the system damping ratio and the undamped natural frequency?

$$G(s) = \frac{3}{s^2 + 7s + 3}$$

7.13. The following plant is to be controlled by a P.D. control system. The steady-state error is to be 5% and the settling time 0.5 second. Calculate the gains required.

$$G(s) = \frac{2}{s^2 + 6s + 5}$$

7.14. The following plant is to be controlled by a P.I.D. control system. Figure P7.14 provides the system pole locations. Calculate the required gains.

$$G(s) = \frac{3}{s^2 + 5s + 4}$$

If only the derivative gain were increased, what would be the effect on system rise time?

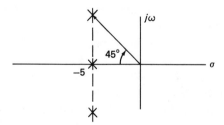

Figure P7.14 Pole locations for Problem 7.14

7.15. Repeat Problem 7.14 if the poles are to be located as shown in Fig. P7.15.

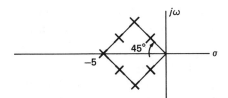

Figure P7.15 Pole locations for Problem 7.15

7.16. Repeat Problem 7.14 if the system pole locations are as shown in Fig. P7.16.

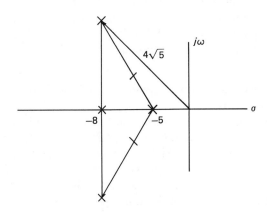

Figure P7.16 Pole locations for Problem 7.16

7.17. The plant in Problem 7.14 is to be controlled by a P.I. control system. If the embedded second-order system is to have a damping ratio of 0.707, calculate the gains required for an optimal system. Plot the pole-zero pattern.

7.18. The plant in Problem 7.14 is to be controlled by a P.I.D. control system. The first-order system is to have a 1-second settling time, and the embedded second-order system is to have a damping ratio of 0.707. Calculate the gains required for an optimal system.

7.19. Plot the responses of Problems 7.14–7.18. Which do you feel to be the best? Is this a valid comparison?

7.20. If you refer back to Problem 5.10, we provided an analog simulation for a proportionally controlled control system. Figure P5.10 provided the analog circuit. If we replace the gain block with our new blocks in this chapter, an analog simulation of P.I., P.D., P.I.D. control is possible. Let's investigate these new types of control on the motor given in Problem 5.10. In all cases:
(a) Calculate the gains required.
(b) Mathematically calculate the steady state-error, the system output, and rise time for a step input ($R(s) = 1/s$). Plot the system output.
(c) Calculate the component values for circuits to satisfy gains calculated in (a).

(d) Build and test the simulation of the control system. Obtain the responses and control action and compare to the theoretical values. Why is the control action noisy in some cases?

PART A P.D. control simulation of plant in Problem 5.10.
The speed-control system is to have a 2-second settling time and a 0.707 damping ratio. Use the circuit in Fig. 7.16 to provide the P.D. block and do parts (a) to (d) above.

PART B P.I. control simulation of plant in Problem 5.10.
The speed-control system is to have a 2-second settling time for the first-order system and a damping ratio of 0.707 for the embedded second-order system. Use the circuit in Fig. 7.20 to implement the P.I. block and do parts (a) to (d) above.

PART C P.I.D. control simulation of plant in Problem 5.10.
The speed-control system is to have a 2-second settling time for the first-order system and a 0.707 damping ratio for the embedded second-order system. Let $K_d = 2$. Use the circuits in Fig. 7.24 to implement the P.I.D. block. If you use circuit (b), $R = 1$ MΩ and feeds directly into the inverting input of the operational amplifier from your simulation of $G(s)$. Is the control action measurable directly in this circuit? Why? Do parts (a) to (d) above. Provide a pole-zero plot for this control system. Generate an optimal P.I.D. control system based on the specifications given. Plot a pole-zero pattern for this system. How does it compare with the original P.I.D. control system? Investigate the response of the speed-control system if only one of the gains is doubled in the original P.I.D. control system, all the other gains remaining the same. Comment on the accuracy of the analog simulation.

Continuous

8

Design and Analysis of Continuous Control Systems

8.1 INTRODUCTION

In this chapter we will investigate the design and analysis of control systems by investigating the migration of the closed-loop poles; this method is the *root-locus* approach to system design and analysis. We then investigate the open-loop frequency response as a tool for design and analysis; this method is the *Bode-plot* approach to system design and analysis.

We look at popular compensating networks and analyze the effects of their placement in the closed loop. The networks include lead, lag, and lead-lag functions. Derivative (rate) feedback and feedforward mechanisms also are studied.

Let's begin with the root-locus approach.

8.2 ROOT-LOCUS ANALYSIS AND DESIGN

This method provides a graphical means of showing the migration of the closed-loop poles as the loop gain varies from zero to infinity. In fact this is a plot

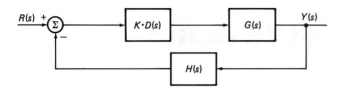

Figure 8.1   Closed-loop control system

of the closed-loop characteristic equation for all values of gain.   Consider the control system in Fig. 8.1.   The closed-loop characteristic equation is given by

$$1 + KD(s)G(s)H(s) = 0 \qquad (8.2.1)$$

Equation (8.2.1) can be put into the following form reflecting the poles and zeroes of $D(s)G(s)H(s)$:

$$1 + K \frac{\Pi_{j=1}^{N}\, (s + z_j)}{\Pi_{i=1}^{M}\, (s + p_i)} \qquad (8.2.2)$$

where M = number of poles, N = number of zeroes, and $-p_i$, $-z_j$ are the pole and zero locations, respectively.

Rewriting equation (8.2.2), we have

$$\Pi_{i=1}^{M}\, (s + p_i) + K\,\Pi_{j=1}^{N}\, (s + z_j) = 0 \qquad (8.2.3)$$

From equation (8.2.3) it is evident that if $K = 0$, the locus begins with the poles, and if $K = \infty$, the locus ends at the zeroes of the transfer function $D(s)G(s)H(s)$.   This transfer function is known as the **open-loop transfer function**.   Let's look at the rules for generating the root locus.

Rule 1.   The first thing to do is to generate the open-loop transfer function.   The pole-zero plot is generated from this transfer function.   As the open-loop poles tend toward the open-loop zeroes, we must have an equal number of poles and zeroes.   The transfer function provides the finite zeroes, and the balance of the zeroes are considered to be at infinity.

Rule 2.   The locus includes all points along the real axis of the s-plane that are to the left of an odd number of poles and finite zeroes.   See Fig. 8.2.

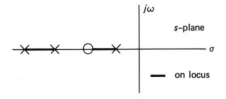

Figure 8.2   Root locus on real axis

Rule 3.   As the gain K approaches infinity, the branches of the locus become asymptotic to straight lines with angles given by ψ:

$$\psi = \frac{180° + 360°k}{M - N}; \quad k = 0, 1, 2 \ldots \tag{8.2.4}$$

Rule 4.   The starting point on the real axis from which the asymptotic lines emanate is given by (σ_c), the centroid of the locus.

$$\sigma_c = \frac{\sum\limits_{i=1}^{M} (-p_i) - \sum\limits_{j=1}^{N} (-z_j)}{M - N} \tag{8.2.5}$$

Rule 5.   If a segment of the locus on the real axis is bounded by two poles, then a breakaway point exists. See Fig. 8.3(a). If a segment of the locus on the real axis is bounded by two zeroes, then a breakin point exists. See Fig. 8.3(b).

If c represents the breakaway or breakin point, then we have

$$\sum\limits_{i=1}^{M} \frac{1}{c + p_i} = \sum\limits_{j=1}^{N} \frac{1}{c + z_j} \tag{8.2.6}$$

In solving equation (8.2.6) it will become evident that a unique solution exists. If more than one value is correct for a breakaway or breakin point, then a mathematical error has been made.

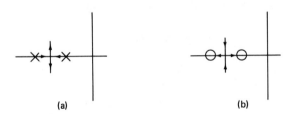

(a)   (b)

Figure 8.3 Root-locus (a) breakaway point, (b) breakin point

EXAMPLE 8.1

Given the control system in Fig. 8.4, provide a complete root locus for this system and indicate system response as the gain increases.

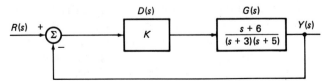

Figure 8.4 Control system for Example 8.1

SOLUTION   We begin by generating the open-loop transfer function.

$$D(s)G(s)H(s) = K \frac{s + 6}{(s + 3)(s + 5)}$$

From the open-loop transfer function we obtain the following information:

$$M = 2, \quad N = 1, \quad p_1 = 3, \quad p_2 = 5, \quad z_1 = 6$$

with the pole locations being at -3, -5 and the zero location at -6. If we plot the poles and zero and apply rule 2, we obtain the locus as shown in Fig. 8.5(a).

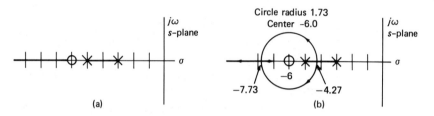

Figure 8.5   Root locus (a) applying rules 1 and 2, (b) completed

If we apply rule 3, the asymptotic lines have an angle of 180°, and by applying rule 4 we find that they emanate from the centroid at -2. This implies that our root locus eventually tracks the real axis and the response is strictly damped exponential.

By applying rule 5 we generate the following equation:

$$\frac{1}{c + 3} + \frac{1}{c + 5} = \frac{1}{c + 6}$$

and solving yields two values ($c_1 = -4.27$, $c_2 = -7.73$). From Fig. 8.5(a) we find the first value to be a breakout point and the second a breakin point. We find the breakin and breakout points to be equidistant from the finite zero location. This implies a circular path for the locus. Figure 8.5(b) provides the completed root locus for the control system.

From Fig. 8.5(b) we find the system to have two poles and one finite zero. The remaining zero is at infinity. Up until now we have not mentioned explicitly the role of the finite zero. We see the finite zero forcing the system poles back to the real axis. This implies that our system has a minimum damping ratio and rise time at the tangent to the circle from the origin and breakin point, respectively. As the gain increases beyond the breakin point, we find one pole heading toward the origin of the s-plane to meet the finite zero at -6. This pole becomes dominant and our system slows down with an increase in gain. It is evident that this system will never become unstable.

This type of analysis is helpful in the design process. The root locus provides an abundance of information and allows the design engineer the graphical means to modify his design. Let's see what happens to a system with no finite zeroes.

EXAMPLE 8.2

Consider the control system given in Fig. 8.6. Provide a complete root locus for this system and find the maximum gain allowed in the loop. An analysis of system response with an increase in gain is also required.

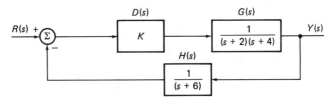

Figure 8.6 Control system for Example 8.2

SOLUTION   The open-loop transfer function is given by

$$D(s)G(s)H(s) = K \frac{1}{(s + 2)(s + 4)(s + 6)}$$

From the open-loop transfer function we obtain the following information:

$$M = 3, \quad N = 0, \quad p_1 = 2, \quad p_2 = 4, \quad p_3 = 6$$

with the pole locations being at -2, -4, -6. There are no finite zeroes, which implies there are three at infinity. If we plot the poles and apply rule 2, we obtain the locus as shown in Fig. 8.7(a).

   If we apply rule 3, the asymptotic lines have angles of 60°, 180°, 300°, and by applying rule 4 we find that they emanate from the centroid at -4. Applying rule 5, we generate the following equation:

$$\frac{1}{c + 2} + \frac{1}{c + 4} + \frac{1}{c + 6} = 0$$

and solving yields two values ($c_1 = -2.85$, $c_2 = -5.15$). From Fig. 8.7(a) we find the first value to be the breakaway point, while the second is not on the locus and is discarded. Figure 8.7(b) provides the completed locus for the control system.

   To find the maximum gain for the system we turn our attention to the characteristic equation. From Fig. 8.6 the characteristic equation is given by

$$s^3 + 12s^2 + 44s + (48 + K) = 0$$

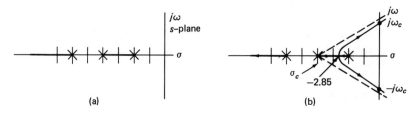

Figure 8.7   Root locus (a) applying rules 1 and 2, (b) completed

From Fig. 8.7(b) we find the locus crosses at $s = j\omega_c$, and substituting into the characteristic equation we have

$$(48 + K - 12\omega_c^2) + j(44\omega_c - \omega_c^3) = 0$$

Setting both the real and imaginary parts to zero, we obtain $\omega_c = 6.63$ rad/sec and $K = 480$. Therefore the maximum gain for the system is 480.

If we study Fig. 8.7(b), we find that our system begins as a damped exponential type of response. As the gain increases, we find one pole continuing to infinity on the negative real axis. This implies the other two poles become dominant and our system behaves much like a second-order system. In that case we could mathematically calculate the gain, given a certain damping ratio. This can be done graphically from an accurate root locus, as the damping ratio is the cosine of the angle made by joining the origin and a point on the root locus. It is interesting to note that our system can become unstable. This is a result of the third pole and no finite zeroes.

All our examples thus far have had poles and zeroes on the negative real axis when the gain was zero. What happens if the system poles or zeroes begin above the negative real axis? Let's look at rule 6.

Rule 6.   The angle of departure from a complex pole or the angle of arrival at a complex zero can be determined as follows. Sum all the angles made when lines are drawn from the complex pole or zero to all the remaining poles and zeroes, and then add 180° to the sum. Angles to similar entities are considered negative; conversely, angles to dissimilar entities are positive. Figure 8.8 provides the angle of departure (λ) for complex pole P_1.

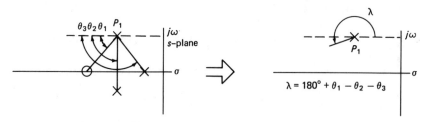

Figure 8.8   Angle of departure for complex pole

EXAMPLE 8.3

The following is the open-loop transfer function for a control system. The system is to be operated at a unity damping ratio. Calculate the required gain K.

$$D(s)G(s)H(s) = \frac{20K(s + 2)}{(s + 4)^2 + 9}$$

SOLUTION   From the open-loop transfer function we have

$$M = 2, \quad N = 1, \quad p_1 = 4 - j3, \quad p_2 = 4 + j3, \quad z_1 = 2$$

with the pole locations being at $-4 \pm j3$ and the zero location at -2. Since the poles are complex, we can apply rule 6 to obtain the departure angle λ. From Fig. 8.9 we have the departure angle being

$$\lambda = 180° - \theta_1 + \theta_2 = 213.7°$$

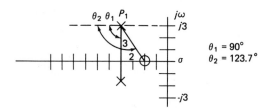

$\theta_1 = 90°$
$\theta_2 = 123.7°$

Figure 8.9   Departure angle for Example 8.3

This implies that our locus is heading back to the negative real axis. Using rule 5, we obtain the breakin point as -5.61. Figure 8.10 provides the completed root locus for the system.

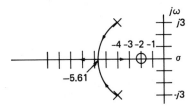

Figure 8.10   Complete root locus for Example 8.3

From Fig. 8.10 we find the system having unity damping ratio at the breakin point. If we substitute $s = -5.61$ into the closed-loop characteristic equation, we can solve for the required gain. In doing so we obtain $K = 0.16$.

If we examine the previous three examples, we find that our analysis was design oriented. We investigated the time response of the control system as the gain was increased. We indicated whether the system would have any stability problems, and, given system specifications (damping ratio), we calculated the required gain. The area of compensation design is highly complex and is a function of the specifications given for a particular control system. It would be a fruitless endeavor to generate a rigorous mathematical menu for compensation design. Instead, let's use the root locus to understand the effects on system response, given a particular compensation network.

EXAMPLE 8.4

Consider a DC motor being used to perform a high-speed positioning function. The transfer function for the motor is second order and relates the output shaft speed to armature voltage. If we integrate the output of the motor, we obtain distance or position. A common compensation network incorporates the addition of a pole and zero with adjustable gain into our closed loop. This control system is shown in Fig. 8.11 with appropriate numerical constants. The zero location is left as a constant a, and its relative location is to be decided upon by the following criteria: the positioning loop should exhibit high speed with zero overshoot; the dominant poles should have a unit damping ratio.

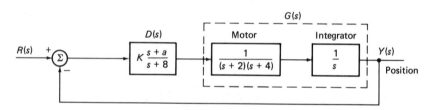

Figure 8.11   Control system for Example 8.4

Since the pole location of the compensation network is fixed, we can explore moving the zero via root locus. To prevent overshoot, the dominant system poles must remain on the negative real axis. If we consider zero locations $a = -10, -6, -3, -1$ and generate a root locus for each case, we will be able to choose the best form of compensating network. Figure 8.12(a)–(d) provides the respective root loci.

It is evident that the zero at -1 would be a bad choice, as the pole at the origin would be the dominant one. This translates to a sluggish system. In practice the zero would be placed at least two poles from the origin to get away from this problem. If we look at the other three options, we find little difference in terms of the root loci. It is apparent that in all three remaining cases the quickest system would be attained if the dominant poles had a breakaway point of -1. If we calculate the closed-loop transfer function, we obtain

$$s^4 + 14s^3 + 56s^2 + (64 + K)s + K \cdot a = 0 \qquad (1)$$

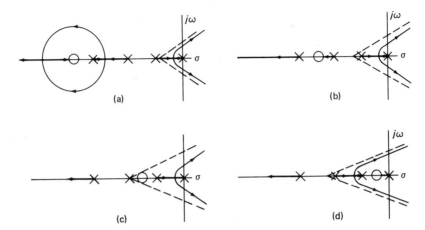

Figure 8.12   Root locus for (a) $a = -10$, (b) $a = -6$, (c) $a = -3$, (d) $a = -1$

If there is to be a breakaway point at $s = -1$, we have

$$(s + 1)(s + 1)$$

as a factor for equation (1). If we perform long division and set the remainder to zero, we arrive at $K = 10$, $a = 3.1$. Therefore Fig. 8.12(c) would be the representative root locus for our compensated control system.

The preceding example is typical of a root locus design. The compensating network is injected and possible root loci are generated to fulfill system specifications. Once the graphical placement of poles is decided upon, the mathematical exercise begins to accurately calculate the unknown variables. The reader is reminded of the importance of understanding the basics, as the root locus is a mere application of the same. It should be remembered that the art of design is to understand fully the system that you are to control. The understanding is aided by mathematics, while experience provides the artistry.

In case you are curious, the compensating network that we created is a lead type compensator, which will be covered later on in the chapter. First, however, let's investigate analysis and design via frequency response, or Bode analysis and design.

8.3 BODE ANALYSIS AND DESIGN

In Section 4.5 we investigated the stability of a system relative to the frequency response. It was evident that another $-180°$ phase shift from our compensation network, plant and transducer would cause positive feedback or instability.

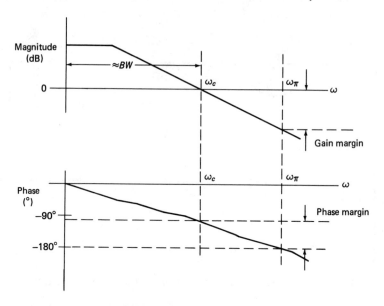

Figure 8.13   Frequency-response curves (magnitude and phase)

This implies that the open-loop transfer function must have less than $-180°$ phase shift to promote stability.   It was also evident that a minimum gain of unity had to exist in the loop for the system to become unstable.   This information is shown in Fig. 8.13.

Since transfer functions are composed of a ratio of factors, it is convenient to use a logarithmic scale (base 10) for the magnitude and frequency axes. The standard Bode plot is attained using semi-log paper.   In this case the frequency is plotted directly, while the magnitude is mathematically calculated as follows:

$$dB \overset{\triangle}{=} 20 \log [\text{magnitude}] \qquad (8.3.1)$$

The phase is always plotted in degrees.

From Fig. 8.13 we also have new terms associated with the frequency-response curves.

Crossover frequency (ω_c)   *As indicated in Chapter 4, this represents the frequency at which the open loop-gain is equal to unity (0 dB). Sometimes known as **gain crossover frequency.***

Bandwidth (BW)   *This is approximately equal to ω_c.   The actual value is higher, as the bandwidth is by definition the frequency at the -3 dB point.*

Phase crossover frequency (ω_π)   *The frequency at which the phase shift is $-180°$.*

TABLE 8.1 Common Factors in Transfer Functions

Factor	s-Plane implication
1. $\dfrac{1}{(j\omega)^k}$, $k = 1, 2$	Multiple pole at origin
2. $(j\omega)^k$, $k = 1, 2$	Multiple zero at origin
3. $\dfrac{1}{1 + j\dfrac{\omega}{a}}$	Single pole at $-a$
4. $1 + j\dfrac{\omega}{b}$	Single zero at $-b$

Gain margin (GM)   *The gain in decibels between the gain and phase crossover frequencies.   This is considered a system design specification.*

Phase margin (PM)   *The angle in degrees between the gain and phase crossover frequencies.   This is considered a system design specification.*

In this day and age there are many computer programs that provide Bode plots accurately, and the reader may like to develop his or her own. As a preliminary paper design it would be nice to quickly sketch the Bode plots for a particular control system. This approximation for the Bode analysis will provide an estimate and appreciation for system stability. As in the case of root locus we are looking for understanding and interpretation of the control system in question. Let's develop the approximations of the Bode plots for the common factors found in transfer functions.

Table 8.1 represents the most common factors found in transfer functions. If we generate the approximate Bode plots for these factors, then any transfer function can be attained by graphically combining the individual components. This is possible because the plots are logarithmic and obey the laws of logarithms. Figures 8.14(a) through (d) provide the approximate Bode plots for factors 1 to 4, respectively. The reader is asked to generate the plots via computer and compare with the approximations. The frequency response of a transfer function is covered in Chapter 3, and the decibel (dB) conversion is given by equation (8.3.1).

Before we tackle a Bode plot for a control system, let's provide some systematic steps for the endeavor.

Step 1.   In order that we may use the plots in Fig. 8.14, we have to put

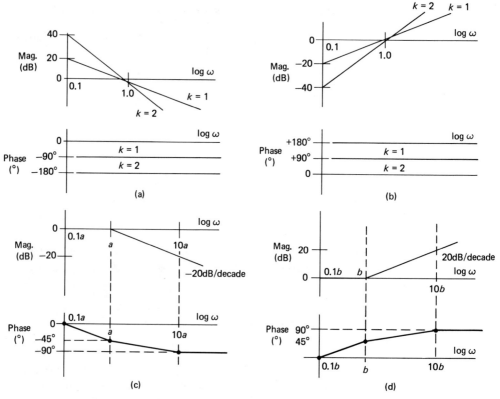

Figure 8.14   Bode plots for factors: (a) 1, (b) 2, (c) 3, (d) 4

our open-loop transfer function in proper Bode form.   That is,

$$D(j\omega)G(j\omega)H(j\omega) = K_B \frac{\left(1 + j\dfrac{\omega}{z_1}\right)\left(1 + j\dfrac{\omega}{z_2}\right) \cdots \left(1 + j\dfrac{\omega}{z_N}\right)}{\left(1 + j\dfrac{\omega}{p_1}\right)\left(1 + j\dfrac{\omega}{p_2}\right) \cdots \left(1 + j\dfrac{\omega}{p_M}\right)} \qquad (8.3.2)$$

where K_B is the Bode gain.

We proceed to plot the Bode gain (in decibels) on the magnitude plot, and this will be the reference line.

EXAMPLE 8.5

The following represents an open-loop transfer function for a control system. Put the transfer function into Bode form to facilitate an approximate Bode plot.

$$D(s)G(s)H(s) = \frac{60(s + 1)}{(s + 2)(s + 3)}$$

SOLUTION   By applying equation (8.3.2) to the open-loop transfer function we obtain, after simplification,

$$D(j\omega)G(j\omega)H(j\omega) = 10 \frac{\left(1 + j\dfrac{\omega}{1}\right)}{\left(1 + j\dfrac{\omega}{2}\right)\left(1 + j\dfrac{\omega}{3}\right)}$$

and, applying equation (8.3.3) to the Bode gain, we obtain $K_B = 20$ dB.

Step 2.   Plot all the individual contributions to magnitude from the poles and zeroes using the Bode gain as a reference line.   Figure 8.14 will provide the magnitude approximations for each pole and zero.

Step 3.   Plot all the individual contributions to phase from the poles and zeroes using Fig. 8.14.   From Fig. 8.14 we find the phase being zero until $\frac{1}{10}$ of the break frequency and at full value at 10 times the break frequency.   The break frequency is the pole or zero frequency in question.   The phase plot is accurate at the break frequency, while the magnitude plot is 3 dB high.

Step 4.   Add all the contributions in both the magnitude and phase plots. The resultant curves will be the approximate response curves for the control system.

EXAMPLE 8.6

Using the open-loop transfer function given in Example 8.5, generate a Bode plot (magnitude and phase).   Indicate the bandwidth, gain margin, and phase margin of the system.

SOLUTION   From Example 8.5 we have the Bode form and gain.   Figure 8.15(a) on page 142 provides the individual magnitude contributions and the final magnitude.   Figure 8.15(b) provides the individual phase contributions and the final phase response.

From Fig. 8.15(a) we find a gain crossover frequency of 40 radians and a −3 dB frequency of 57 radians, which represents the bandwidth of the system.   We indicated that the system bandwidth could be approximated by the gain crossover frequency.   In this case the error is large and is the result of a gently sloping magnitude plot.   The reader should be aware of this and use the −3 dB frequency for accurate bandwidth values.   If we perform the necessary mathematics, we find that the bandwidth is 59.9 radians.   From Fig. 8.15(b) we find the phase crossover frequency to be infinite and the gain and phase margin likewise.

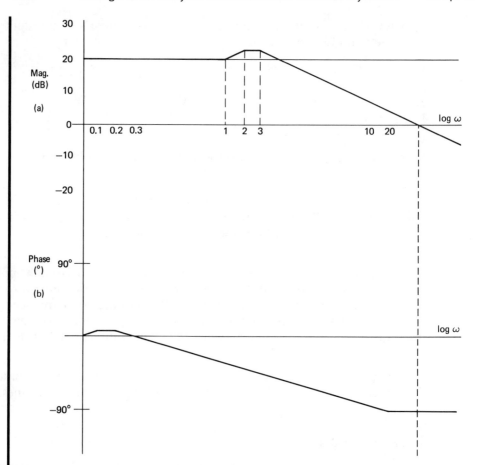

Figure 8.15   Bode plot: (a) magnitude, (b) phase.   Light lines represent individual contributions.

From the Bode plots we find the control system very stable.   One form of compensation is to inject gain into the loop.   This would be reflected in an increase of the Bode gain.   The magnitude plot would shift upward and the phase plot would not change.   The upward shifting of the magnitude plot implies an increase in bandwidth and therefore a quicker system.   If we generate the root locus for our system, we find the poles remaining on the negative real axis for all values of gain.   The dominant pole moves toward the origin to the finite zero at -1.   There seems to be a discrepancy here, as this implies a sluggish system.   The reader is asked to generate the root locus and solve for the system response for various gains.   The coefficients on the dominant exponential term should be compared, as the answer lies within.

The Bode-plot method provides a convenient means of graphically generating a control system to meet specifications. The control system will be specified in terms of maximum gain crossover frequency and a minimum phase and gain margin. We will look at some compensating options quite shortly, but first let's look at a realistic problem.

EXAMPLE 8.7

The control system in Example 8.6 was very stable and exhibited a phase shift of $-90°$ at the gain crossover frequency of 40 radians. The system in question has a mechanical component that exhibits a certain amount of transport lag or dead time. This implies that once an input change is sensed, it waits for a certain time to respond, say T_D seconds. The design engineer is interested to find out the maximum amount of dead time this system can tolerate, as this will reflect the cost of the mechanical component in an inverse sense.

SOLUTION   At first this seems a futile exercise, as we have indicated by root locus and Bode plots that this system will remain stable for all values of gain. If we refer back to the shifting property in Laplace transforms, we can show a transport lag of T_D seconds as

$$e^{-s \cdot T_D}$$

and, finding the frequency response, we have

$$\cos(\omega \cdot T_D) - j \sin(\omega T_D)$$

which in polar form is given by

$$1 \angle -\omega \cdot T_D$$

It is evident that our phase plot will be adjusted. Substituting our gain crossover frequency, converting to degrees, setting equal to the balance of the phase shift ($-90°$), and solving for the dead time, we obtain $T_D = 0.039$ second.

The dead time provides a lot of phase shift and is quite visible with the Bode analysis. The reader should appreciate that a very stable system can become unstable with the proper amount of transport lag. As transport lag is apparent in every physical system, it is important that it be considered in terms of gain reduction or the proper amount of phase and gain margin in the original specification.

We now turn our attention to compensating networks that will allow pole and zero manipulation in the s-plane in terms of the root locus. As for the Bode plots, we will be able to modify bandwidth, gain, and phase margins. Let's begin with the lead compensator.

8.4 LEAD COMPENSATOR

The typical form of a **lead compensator** $D(s)$ is given by

$$D(s) = K \frac{(s + a)}{(s + b)}; \quad b > a \tag{8.4.1}$$

Figure 8.16 provides the appropriate circuit to perform lead compensation.

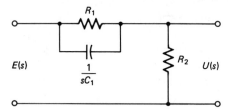

Figure 8.16 Lead compensation network

Finding the transfer function for the circuit in Fig. 8.16, we obtain the constants

$$a = \frac{1}{R_1 C_1} \tag{8.4.2}$$

$$b = a + \frac{1}{R_2 C_1} \tag{8.4.3}$$

$$K = 1 \tag{8.4.4}$$

From equation (8.4.1) we find the network attenuating, and to attain unity DC gain the loop gain must be increased by b/a. If the pole location is much greater than the zero location, we find the lead compensator approximating a pure derivative function. In Chapter 7 we found that these functions are susceptible to noise, and care should be taken in their application.

In Example 8.4 we examined the application of a lead compensator using the root locus. We indicated that the zero should be located to the left of at least two dominant poles. In Fig. 8.17(b) we see the result of a lead compensator on a second-order system. Comparing with the uncompensated system [Fig. 8.17(a)] we find an increase in system speed with the compensated

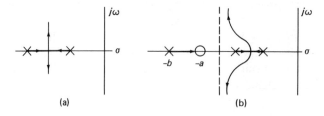

Figure 8.17 Second-order system: (a) alone, (b) with lead compensation

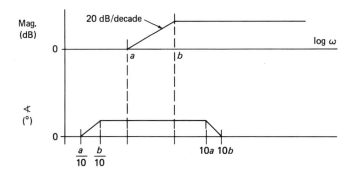

Figure 8.18   Frequency response for lead compensator

system.   What about the frequency response?   Let's look at the Bode plots for the lead compensator.

Figure 8.18 provides the Bode plots for the lead compensator.   From Fig. 8.18 we find that this type of compensation network can increase bandwidth and also phase margin.   If we take equation (8.4.1) and find the maximum phase shift, it occurs at frequency ω_ϕ, which is

$$\omega_\phi = \sqrt{a \cdot b} \tag{8.4.5}$$

and the maximum phase shift (ϕ_m) at (ω_ϕ) is

$$\phi_m = \arctan\left[\frac{b - a}{2\sqrt{ab}}\right] \tag{8.4.6}$$

From the circuit in Fig. 8.16 it is a good practice to have the resistors mismatched by no more than a factor of ten.   Practically, equation (8.4.3) would be approximated by

$$b = 11a \tag{8.4.7}$$

If we substitute equation (8.4.7) into equation (8.4.6), we find the maximum phase shift from this network is approximately 60°.   If the compensation network requires more phase lead (positive phase shift), then the practical solution is to cascade two lead compensators.

EXAMPLE 8.8

A control system is operating at an acceptable phase margin of $-60°$.   The gain crossover frequency is at 10 radians.   If the gain is increased such that the gain crossover frequency becomes 15 radians, the system exhibits a very low phase margin.   The design engineer specifies a lead compensator that maintains the original phase margin (60°) at a gain crossover frequency of 15 radians.   If the original system had a phase shift of $-160°$ at 15 radians, calculate the pole and zero locations and possible circuit values.

SOLUTION From equation (8.4.5) we have

$$15 = \sqrt{a \cdot b} \tag{1}$$

Now we require our maximum phase lead at our new gain crossover frequency. To obtain a phase margin of 60° we require a phase lead of 40°. Substituting into equation (8.4.6), we have

$$40° = \arctan \left[\frac{b - a}{2\sqrt{ab}} \right] \tag{2}$$

If we substitute equation (1) into (2) and solve, we obtain $a = 7$ and $b = 32$. If we use a 1μF capacitor for C_1, we obtain $R_1 = 143$ KΩ and $R_2 = 40$ KΩ with respect to the circuit in Fig. 8.16. To offset the attenuation due to the passive network we would have to increase our gain by a factor of 4.6, as indicated by the attenuation factor b/a mentioned before. Therefore the required compensation network $D(s)$ would be given by

$$D(s) = 4.6 \frac{s + 7}{s + 32}$$

As stated earlier, the lead compensator will increase the speed of response for a system and also increase the phase margin. In some cases it is necessary to slow a system down and also maintain the phase margin. In this case a lag compensator would be used.

8.5 LAG COMPENSATOR

The typical form of a **lag compensator** $D(s)$ is given by

$$D(s) = K \frac{s + a}{s + b}, \qquad a > b \tag{8.5.1}$$

Figure 8.19 provides the appropriate circuit to perform the lag compensation. Finding the transfer function for the circuit in Fig. 8.19, we obtain the following constants:

$$a = \frac{1}{R_2 C_2} \tag{8.5.2}$$

$$b = \frac{1}{(R_1 + R_2)C_2} \tag{8.5.3}$$

$$K = \frac{b}{a} \tag{8.5.4}$$

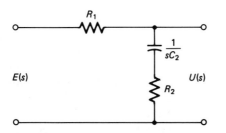

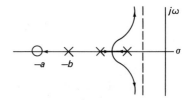

Figure 8.19 Lag compensation network

Figure 8.20 Lag compensation of second-order system

In this case we find the dc gain to be unity, and no gain factor is required as in the case of the lead compensator. If the pole location is very small, we find the lag compensator approximating a pure integrator. This implies less steady-state error, one of the major uses for the lag compensator.

In Fig. 8.20 we see the result of adding a lag compensator to a second-order system as shown in Fig. 8.17(a). We find the dominant poles being shifted toward the $j\omega$ axis. This implies a slower system with less error. Let's investigate the frequency response.

Figure 8.21 provides the Bode plots for the lag compensator. From Fig. 8.21 we find that this type of compensation can decrease bandwidth and, unfortunately, phase margin. If we guarantee that the zero is $\frac{1}{10}$ of the required gain crossover frequency, then our phase margin will stay the same. From our uncompensated system we can graphically obtain the surplus of gain at the new crossover frequency (K_c) and set this equal to the gain loss of our lag network. Mathematically we can show these as

$$a = \frac{\omega_c}{10} \tag{8.5.5}$$

$$K_c = 20 \log \left(\frac{a}{b} \right) \tag{8.5.6}$$

Equations (8.5.5) and (8.5.6) can be used to quickly design a lag compensator for a control system. Let's look at an example.

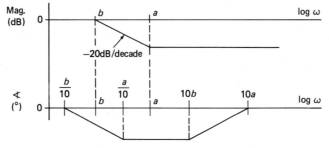

Figure 8.21 Frequency response for lag compensator

EXAMPLE 8.9

A control system is operating with a gain crossover frequency of 18 radians. The control engineer would like to decrease the gain crossover frequency to 12 radians. The phase margin of the original system at 12 radians is 45° and considered to be satisfactory. The surplus gain (K_c) at 12 radians of the original system is 25 dB. Find the location of the zero and pole for a lag compensator to provide these specifications. The circuit values are also required to implement the compensator.

SOLUTION   From equation (8.5.5) we have

$$a = 1.2 \qquad (1)$$

Substituting equation (1) into equation (8.5.6) and solving, we obtain $b = 0.067$. If we assume a 4.7-μF capacitor, we obtain $R_2 = 180$ KΩ and $R_1 = 3.0$ MΩ from equations (8.5.2) and (8.5.3), respectively. The lag compensator $D(s)$ is given by

$$D(s) = 0.056 \frac{s + 1.2}{s + 0.067}$$

In some control systems the control engineer may require a combination of the lead and lag functions. It is very easy to cascade a lead and lag function, as we have discussed in this and the preceding section. If the control engineer requires a single network to perform the dual purpose, then a lead-lag network is required. Let's briefly look at this compensating network.

8.6 LEAD-LAG COMPENSATION

A typical form of a lead-lag compensator $D(s)$ is given by

$$D(s) = K \frac{(s + a_1)(s + a_2)}{(s + b_1)(s + b_2)}; \quad a_1 < b_1, a_2 > b_2 \qquad (8.6.1)$$

Figure 8.22 provides the appropriate circuit to perform the lead-lag compensation. Solving for the transfer function for the circuit in Fig. 8.22, we obtain the following constants:

$$a_1 = \frac{1}{R_1 C_1} \qquad (8.6.2)$$

$$a_2 = \frac{1}{R_2 C_2} \qquad (8.6.3)$$

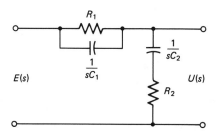

Figure 8.22 Lead-lag compensation network

$$a_1 a_2 = b_1 b_2 \tag{8.6.4}$$

$$a_2 + b_2 = a_1 + b_1 + \frac{1}{R_2 C_1} \tag{8.6.5}$$

$$K = 1 \tag{8.6.6}$$

The DC gain for this compensator is unity, and no further gain factor is required. As the equations indicate, this compensating network appears to be more difficult to apply. As this application can be attained by a lead and lag separately, we will not spend any time on it. The reader should understand that the application for this type of compensator would be called upon if the system required attenuation and an increase in phase margin.

In all our examples it was assumed that the control systems were operating with a suitable gain and that our compensating networks would provide the change in gain and phase margin required. The preceding introduction to the design of compensating networks is by no means complete. It is intended as a primer for the vast field of compensation design.

We now leave our compensating networks to look at two popular techniques used in control systems, the first being derivative or rate feedback.

8.7 DERIVATIVE FEEDBACK

Derivative or **rate feedback** is a common practice in servo control systems. In fact it is applicable to any system that has a signal available that is equivalent to the derivative of the controlled variable. Figure 8.23 provides the block diagram for a derivative feedback scheme for a second-order plant. The gain K represents the forward gain factor, and K_f represents the derivative feedback gain factor. In servo systems it is desirable to have high forward gain factors K to facilitate a minimal deadband about the position setpoint. Without derivative feedback this high forward gain would promote high oscillations about the setpoint. This property is not the norm for an accurate high-speed posi-

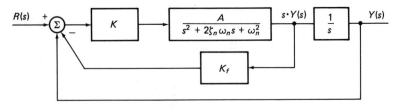

Figure 8.23   Control system with derivative feedback

tioning system, generally known as a *motion control system*.   Let's investigate
the properties of our system under derivative feedback.

From Figure 8.23 we find the overall transfer function $P(s)$ as

$$P(s) = \frac{K \cdot A}{s^3 + 2\zeta_n\omega_n s^2 + (\omega_n^2 + K \cdot K_f \cdot A)s + K \cdot A} \qquad (8.7.1)$$

If we assume $(s + a)$ to be the embedded first-order system, we obtain the
second-order system by long division:

$$s^2 + (2\zeta_n\omega_n - a)s + \frac{K}{a} \qquad (8.7.2)$$

where

$$K = \frac{a\omega_n^2 + a^3 - 2\zeta_n\omega_n a^2}{A(1 - a \cdot K_f)} \qquad (8.7.3)$$

If K is to be large, we have from (8.7.3)

$$1 - a \cdot K_f = 0$$

which simplifies to

$$a = \frac{1}{K_f} \qquad (8.7.4)$$

From equation (8.7.4) it is evident that we can move the first-order system to
the left on the negative real axis.   This can be accomplished by having the
feedback gain factor (K_f) between 0 and 1.   If we want this pole to be dominant,
we have to guarantee that the real part of the second-order system is greater
than the first-order system and equal when $K = \infty$.   That is,

$$\frac{2\zeta_n\omega_n - a}{2} = a$$

which simplifies to

$$a = \frac{2\zeta_n\omega_n}{3} \qquad (8.7.5)$$

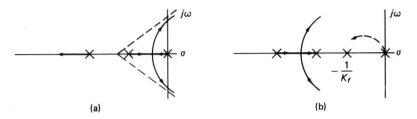

Figure 8.24   Root locus (a) without (b) with derivative feedback

and, using equation (8.7.4), we obtain

$$K_f = \frac{3}{2\zeta_n \omega_n} \tag{8.7.6}$$

The reader is asked to compare equation (8.7.5) to equation (7.6.4).

To appreciate the effects of derivative control consider Fig. 8.24(a) and (b), the root locus for the control system without and with derivative feedback, respectively. From Fig. 8.24(a) we find the system becoming unstable as the gain K is increased. This is a result of the pole at the origin. From Fig. 8.24(b) we find the pole at the origin moving as indicated by equation (8.7.4). This implies that our dominant pole can be first order and still provide a quick response. The derivative feedback provides a shifting of the origin as viewed from the root locus of the control system with no derivative feedback. As long as our dominant pole is first order, we can keep our overshoot minimal due to the embedded second-order system. Equations (8.7.5) and (8.7.6) provide limits as our gain approaches infinity.

An important observation from the practical side pertains to our application of derivative gain in our discussion of P.I.D. control. In that form of control we found derivative control very susceptible to noise and its application difficult. In derivative feedback we find that its effect is maximum when its gain is minimum [equation (8.7.4)]. This implies fewer problems with noise and a practical application for derivative type gain.

Another helpful control scheme is the incorporation of feedforward type control. Let's consider this type.

8.8 FEEDFORWARD COMPENSATION

Feedforward compensation can be utilized to attain a finite zero in a control system with no finite zeroes. Consider the control system in Fig. 8.25. In this case the compensating network is separated and a portion $D_1(s)$ is sent

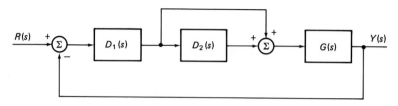

Figure 8.25 Feedforward compensation

forward to the plant. The transfer function for this system is given by

$$\frac{Y(s)}{R(s)} = \frac{D_1(s)G(s)\,[1 + D_2(s)]}{1 + D_1(s)G(s)\,[1 + D_2(s)]} \qquad (8.8.1)$$

From equation (8.8.1) we find the factor $1 + D_2(s)$ implying a finite zero if $D_2(s)$ has poles only. Let's consider an example.

EXAMPLE 8.10

A first-order plant is to be controlled by a P.I. control scheme. After careful analysis it is found that this type of scheme would not fulfill the speed requirements as outlined in the specifications. It is recommended that a feedforward mechanism be investigated, as the system requires zero steady-state error. Figure 8.26 provides the feedforward control of the system.

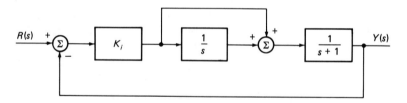

Figure 8.26 Feedforward control of first-order system

From Fig. 8.26 we find that integral control is used to promote zero steady-state error. The integral gain is separated and fed forward to our plant. This can be represented as the derivative of the component added to the plant by the integrating block. From equation (8.8.1) we obtain the system transfer function

$$\frac{Y(s)}{R(s)} = \frac{K_i}{s + K_i}$$

It is evident that our system exhibits zero steady-state error and at the same time remains a first-order system with a speed increase directly related to integral gain K_i. The reader should investigate the response of this system with no feedforward compensation. Will it oscillate?

8.9 SUMMARY

In this chapter we investigated the methods of control system analysis and design. We tracked the closed-loop system poles as a method of analysis. It was evident that a particular location on this locus dictated a particular response. This could be used for the design of a compensating network for a control system. The process was known as root locus.

We turned our attention to the frequency response of a control system and found this to be another vehicle for analysis and design. The gain- and phase-margin approach provided intelligent means of system analysis and design.

As design implies compensation, five types of compensation techniques were investigated. The lead compensator provided a phase lead and gain increase. It provided a derivative type function and increased the system error. The lag compensator provided phase lag and gain decrease. It provided an integral type function and decreased the system error. The lead-lag compensator provided a means of decreasing system gain and increasing phase margin.

The remaining two compensating techniques involved a feedback mechanism. The derivative feedback technique provided control of system speed and overshoot. This technique was considered important in servo control systems. The feedforward technique split the intended compensating network and fed the portions to the plant separately. This provided an introduction of finite zeroes to a system that would otherwise have none.

Problems

8.1. Given the following open-loop transfer functions, provide an accurate root locus for each and indicate system response for the various values of (**a**).

(**a**) $\dfrac{K(s + a)}{(s + 2)(s + 6)}$; $a = 0, 4, 10$

(**b**) $\dfrac{K(s + a)}{(s + 2)(s + 4)(s + 6)}$; $a = 0, 3, 5, 8$

(**c**) $\dfrac{K(s + 1)(s + a)}{(s + 2)(s + 4)(s + 6)}$; $a = 0, 3, 5, 8$

8.2. Given the following open-loop transfer function:

(**a**) Generate an accurate root locus for the system.

(**b**) Find the maximum gain possible.

(**c**) Find the crossover frequency.

(d) Refer to Problem 5.3.   If this system is operating at a gain $K = 1.2$, what is the gain margin?

$$\frac{20K}{s(s + 1)(s + 3)(s + 5)}$$

8.3. Isolate the dominant system in the root locus generated in Problem 8.2. Graphically calculate the system damping ratio if the system undamped natural frequency is 1 rad/sec.   Chapter 6 will help.   Find the gain margin at this point.

8.4. Given the following open-loop transfer function:
   (a) Generate an accurate root locus.
   (b) Find the gain required for a critically damped system.
   (c) What is the minimum damping ratio?
   (d) What is the undamped natural frequency at this damping ratio?

$$\frac{3K(s + 6)}{s^2 + 4s + 13}$$

8.5. In Chapter 7 we investigated the P.I.D. control scheme.   If we consider P.I. control only, we find this block to have a transfer function

$$D(s) = \frac{K_p\left(s + \dfrac{K_i}{K_p}\right)}{s}$$

This transfer function has the pole fixed and a variable zero.   The zero location is always greater than the pole location, which implies a lag compensator.   Consider the following plant to be controlled by a P.I. control system:

$$G(s) = \frac{2}{s^2 + 4s + 3}$$

If the P.I. control system has unity feedback,
   (a) Consider the best place for the zero using root-locus techniques.
   (b) Compare your answer with that obtained in the optimal system as calculated in Example 7.2.

8.6. The P.D. control scheme can be shown as the following transfer function:

$$D(s) = K_d\left(s + \frac{K_p}{K_d}\right)$$

The pole location is considered at infinity for our pure derivative function. As stated in Problem 7.2, this function entails difficulties and in practical terms the pole is finite.   We will assume a pure derivative function.   Since the zero location is always less than the pole location, a lead compensator is implied. If the plant in Problem 8.5 is controlled by a P.D. control scheme with unity feedback,

(a) Find the best place for the zero using root-locus techniques.

(b) Compare to the location obtained in Example 7.1.

8.7. A P.I.D. control scheme can be shown as the following transfer function:

$$D(s) = K_d \frac{\left(s^2 + \dfrac{K_p}{K_d} s + \dfrac{K_i}{K_d}\right)}{s}$$

This function has a pole at infinity and a pole at the origin. The location of the zeroes is variable and therefore a lead-lag compensator is implied. Consider the plant used in Problem 8.5 with unity feedback:

(a) Consider the best place for the zeroes using root-locus techniques.

(b) Compare your answer with that obtained from the time analysis of P.I.D. control in Chapter 7.

8.8. Provide Bode plots for the following open-loop transfer functions. In each case indicate the gain and phase margin.

(a) $\dfrac{100(s + 2)}{(s + 3)(s + 4)}$

(b) $\dfrac{20}{(s + 1)(s + 3)}$

(c) $\dfrac{20(s + 1)}{(s + 2)(s + 4)(s + 6)}$

8.9. Consider the general second-order plant given by

$$G(s) = \frac{1}{s^2 + 2\zeta_n s + 1}$$

If $\zeta_n = 0.1, 0.2, \ldots, 0.9$:

(a) Use a computer to calculate the magnitude and phase for each case.

(b) Plot the magnitude and phase with respect to frequency. Indicate the bandwidth and resonant frequency in each case. Compare to results obtained by using the equations in Problems 3.10 and 3.9, respectively. These curves can be used to complete our straight-line approximation functions given in Table 8.1. Now we can handle imaginary roots.

8.10. Provide the Bode plot for the P.I. compensator as given in Problem 8.5. Use $K_i = 1$. Compare this to the lag compensator Bode plot given in Fig. 8.21.

8.11. Provide the Bode plot for the P.D. compensator as given in Problem 8.6. Use $K_p = 1$. Compare this to the lead compensator Bode plot given in Fig. 8.18.

8.12. Consider Example 8.5. If this system represents a unity-feedback control system, find the closed-loop transfer function. Provide a Bode plot for this function. This represents the closed-loop Bode plot. What is the bandwidth of this transfer function? How does it compare with the open-loop bandwidth calculated in Example 8.5?

8.13. Given the following open-loop transfer function:

$$D(s)G(s)H(s) = \frac{10}{s^2}$$

(a) Generate the Bode plot for the system.
(b) Generate a lead compensator to provide a gain crossover frequency of 5 radians and a phase margin of 45°.
(c) Provide a circuit to implement the lead compensator.
(d) Provide a Bode plot for the compensated system. Does it satisfy the specifications?

8.14. Redo Problem 8.13 with the following open-loop transfer function:

$$D(s)G(s)H(s) = \frac{20}{(s + 2)s}$$

with the gain crossover frequency at 8 radians.

8.15. Design a lag compensator for the open-loop transfer function given in Problem 8.14. The gain crossover frequency is to be 1.5 radians and the phase margin is to be 45°. Provide a circuit diagram to implement the function. Generate the Bode plot for the compensated system. Have the specifications been met? If not, can an increase in gain accommodate the error?

8.16. Show that the maximum phase lag occurs at a frequency given by $\sqrt{a \cdot b}$ rad/sec, for a lag compensator.

8.17. Consider the following open-loop transfer function:

$$D(s)G(s)H(s) = \frac{1}{s^2 + 0.2s + 1}$$

(a) Generate the Bode plot for the open-loop transfer function. Use Problem 8.9 as a reference.
(b) Design a lag compensator to provide a gain crossover frequency of 1 rad/sec with a phase margin of 90°.
(c) Provide a Bode plot for the compensated system. Have the specifications been met?

8.18. Show that a lead compensator will increase the steady-state error in a unity-feedback control system with a step input.

8.19. Show that a lag compensator will decrease the steady-state error in a unity-feedback control system with a step input.

8.20. Consider the tank level control system shown in Fig. P8.20.
  (a) Provide a Bode plot for the system if $T_D = 0$ sec.
  (b) What is the phase margin?

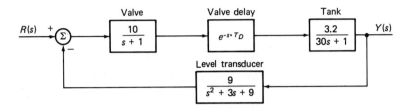

Figure P8.20   Control system for Problem 8.20

  (c) If $T_D = 1.1$ seconds, what is the new phase margin?
  (d) If we require 40° phase margin, calculate the decrease in gain required.

8.21. A lead-lag function is sometimes used to cancel the dominant pole of a plant to increase the control system performance.   Consider the following plant:

$$G(s) = \frac{1}{(s + 0.2)(s + 4)}$$

A lead-lag compensator of the following form is proposed:

$$D(s) = K \frac{(s + 0.2)(s + a)}{s(s + b)}$$

The pole at zero will promote zero steady-state error to a step input, and the zero at -0.2 will cancel the slow pole of the plant.   Use the method of your choice to complete the design of the compensator such that the system will exhibit a damping ratio of 0.707 and an undamped natural frequency of 7.07 rad/sec.

8.22. Given the position control system with unity derivative feedback shown in Fig. P8.22:
  (a) Provide the closed-loop transfer function.
  (b) Show that the first-order system is located at -1 for all values of K.
  (c) Provide the proper derivative feedback and compare the system response with the original system.

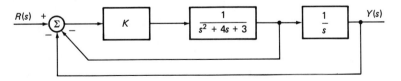

Figure P8.22   Control system for Problem 8.22

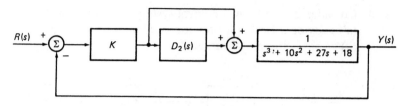

Figure P8.23   Control system for Problem 8.23

8.23. Consider the feedforward control system shown in Fig. P8.23. Find the compensating network $D_2(s)$ and the value of the gain K such that
   (a) The steady-state error for a step input is 10%.
   (b) The damping ratio is 1.01.

Part I
Continuous

9

State-Variable Analysis

9.1 INTRODUCTION

In the preceding chapters we have investigated the control system in the frequency domain. Using the Laplace transform, we converted the system differential equation to an algebraic equation. We developed the transfer function for the control system. Using the transfer function, we investigated the characteristic equation and the importance of system pole locations on the s-plane. We used the frequency domain as a means of analyzing and designing control systems based upon specifications given relative to frequency response. This proved to be a powerful and useful scheme.

As control systems became more complicated, the time-domain analysis of control systems became important. These systems contain multiple inputs and outputs and do not lend themselves to an overall transfer function—a function upon which the entire frequency-domain analysis is based. This time-domain analysis of multivariable control systems is known as **state-variable analysis** and characterizes the era of modern control theory. The reader is reminded that this type of analysis requires knowledge of the frequency-domain analysis as previously covered. The state-variable approach is an extension of control theory to include multivariable, nonlinear, and time-varying control systems. These systems could not be handled with our frequency methods, as we based our analysis on linear time-invariant control systems.

In this chapter we will introduce the system state variables and define the state-variable equations. We will transform our previously defined transfer functions into state-variable form, using signal flow diagrams. To predict stability in the time domain we use our frequency techniques to extract the characteristic equation from the state-variable equations. We introduce two new terms used in modern control theory: controllability and observability.

It is also important for the reader to understand that state-variable design lends itself to adaptive and optimum control design. We will introduce only the analysis portion. The reader who is interested in the state-variable design approach to control systems will find many texts available.

Let's begin with the signal flow diagram and see how our plants defined previously can be shown in the time domain in terms of state variables.

9.2 SIGNAL-FLOW DIAGRAMS

In Chapter 2 we investigated the simulation of physical systems utilizing operational amplifiers. We indicated that the integrating block was the basic building block. The number of integrating blocks in cascade represented the order of the differential equation of the physical system. If we use the same analogy, we can generate a **simulation diagram**, more commonly known as the **phase-variable diagram**. An example will aid the procedure.

EXAMPLE 9.1

Consider a system with the following transfer function:

$$P(s) = \frac{2s^2 + 4s + 6}{2s^3 + 6s^2 + 8s + 10}$$

Provide a phase-variable simulation diagram for the system.

Step 1. Divide all terms by the highest order of the Laplace operator s. This process is valid only for transfer functions with a numerator that is of order less than or equal to the order of the denominator. Performing this on our transfer function, we have

$$P(s) = \frac{\dfrac{1}{s} + \dfrac{2}{s^2} + \dfrac{3}{s^3}}{1 + \dfrac{3}{s} + \dfrac{4}{s^2} + \dfrac{5}{s^3}}$$

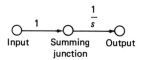

Figure 9.1 Integrator for simulation diagram

Step 2. The highest order of the Laplace operator dictates the number of integrators in our simulation diagram. The integrator is shown in Fig. 9.1. From Fig. 9.1 we see each integrator having an input, output and summing junction. The unity gain reflects the passing of the input to the integrator without modification. In step 2 we cascade the number of integrators reflecting the highest order of the Laplace operator. In our case this is equal to 3. We label each output of the integrator as X_i (s), where i has values from 1 to the number of integrators. The output of the last integrator is indicated as X_1 (s). Figure 9.2 provides step 2 of the simulation diagram for our transfer function.

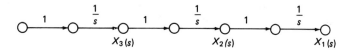

Figure 9.2 Step 2 of simulation diagram

Step 3. Our transfer function $P(s)$ is given by $Y(s)/R(s)$. This implies that our output $Y(s)$ includes feedforward elements given by the numerator of our modified transfer function. The gain for each integrator output is given by the coefficient of the corresponding integrator term of the numerator. Figure 9.3 provides step 3 for the simulation diagram.

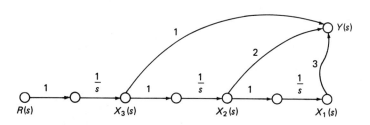

Figure 9.3 Step 3 of simulation diagram

Step 4. If we look at the denominator of the modified transfer function, it is evident that it is in the form $1 + D(s)G(s)H(s)$. This provides the feedback elements in our system. We indicate these feedback elements on our phase-variable simulation diagram as shown in Fig. 9.4. The reader will find the sign on the gain factors reversed, as this reflects the feedback process.

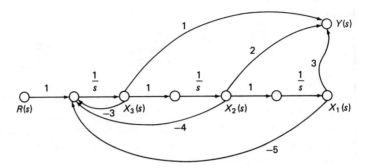

Figure 9.4   Phase-variable simulation diagram for Example 9.1

EXAMPLE 9.2

In Chapter 7 we indicated the past, present, and future sense of control in a P.I.D. control system. Using a phase-variable simulation diagram, demonstrate that this is true for the P.I.D. control system shown in Fig. 9.5.

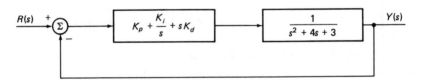

Figure 9.5   P.I.D. controls system

From Fig. 9.5 we obtain the overall system transfer function

$$\frac{Y(s)}{R(s)} = \frac{s^2 K_d + s K_p + K_i}{s^3 + (4 + K_d)s^2 + (3 + K_p)s + K_i}$$

which in modified form is

$$\frac{Y(s)}{R(s)} = \frac{\dfrac{K_d}{s} + \dfrac{K_p}{s^2} + \dfrac{K_i}{s^3}}{1 + \dfrac{4 + K_d}{s} + \dfrac{3 + K_p}{s^2} + \dfrac{K_i}{s^3}}$$

with Fig. 9.6 providing the simulation diagram.

From Fig. 9.6 it is evident that the derivative portion appears at the output before the proportional and integral, respectively. This is true for the feedback contributions. If the proportional contribution is considered

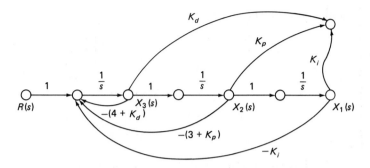

Figure 9.6 Phase simulation diagram for P.I.D. control system

present time, then the derivative and integral are future and past, respectively.

If we refer back to Fig. 9.4, we can generate the following equations:

$$X_1(s) = \frac{1}{s} \cdot X_2(s)$$

$$X_2(s) = \frac{1}{s} \cdot X_3(s)$$

$$X_3(s) = \frac{1}{s} \cdot [-5X_1(s) - 4X_2(s) - 3X_3(s) + R(s)]$$

$$Y(s) = 3X_1(s) + 2X_2(s) + X_3(s)$$

$$s \cdot X_1(s) = X_2(s)$$

$$s \cdot X_2(s) = X_3(s)$$

$$s \cdot X_3(s) = -5X_1(s) - 4X_2(s) - 3X_3(s) + R(s)$$

$$Y(s) = 3X_1(s) + 2X_2(s) + X_3(s)$$

which generates the following time functions:

$$\dot{x}_1 = x_2$$

$$\dot{x}_2 = x_3$$

$$\dot{x}_3 = -5x_1 - 4x_2 - 3x_3 + r(t)$$

$$y(t) = 3x_1 + 2x_2 + x_3$$

and finally, in matrix form:

$$\begin{bmatrix} \dot{x}_1 \\ \dot{x}_2 \\ \dot{x}_3 \end{bmatrix} = \begin{bmatrix} 0 & 1 & 0 \\ 0 & 0 & 1 \\ -5 & -4 & -3 \end{bmatrix} \begin{bmatrix} x_1 \\ x_2 \\ x_3 \end{bmatrix} + \begin{bmatrix} 0 \\ 0 \\ 1 \end{bmatrix} r(t)$$

$$y(t) = \begin{bmatrix} 3 & 2 & 1 \end{bmatrix} \begin{bmatrix} x_1 \\ x_2 \\ x_3 \end{bmatrix}$$

We have just generated the state equations of a control system in terms of the state variables $x_i(t)$. The equations are coupled first-order differential equations with the output equation given by $y(t)$. We now turn our attention to the general form of these equations.

9.3 STATE EQUATIONS

From our phase simulation diagram process any nth-order system with i inputs can be shown as

$$sX_1(s) = a_{11}X_1(s) + a_{12}X_2(s) + \cdots + a_{1n}X_n(s) + b_{11}R_1(s) + \cdots + b_{1i}R_i(s)$$
$$sX_2(s) = a_{21}X_1(s) + a_{22}X_2(s) + \cdots + a_{2n}X_n(s) + b_{21}R_1(s) + \cdots + b_{2i}R_i(s)$$

.

.

.

$$sX_n(s) = a_{n1}X_1(s) + a_{n2}X_2(s) + \cdots + a_{nn}X_n(s) + b_{n1}R_1(s) + \cdots + b_{ni}R_i(s)$$

(9.3.1)

and in the time domain:

$$\dot{x}_1 = a_{11}x_1 + a_{12}x_2 + \cdots + a_{1n}x_n + b_{11}r_1 + \cdots + b_{1i}r_i$$
$$\dot{x}_2 = a_{12}x_1 + a_{22}x_2 + \cdots + a_{2n}x_n + b_{21}r_1 + \cdots + b_{2i}r_i$$

.

.

.

$$x_n = a_{n1}x_1 + a_{n2}x_2 + \cdots + a_{nn}x_n + b_{n1}r_1 + \cdots + b_{ni}r_i$$

(9.3.2)

which in matrix form is

$$
\begin{bmatrix} \dot{x}_1 \\ \dot{x}_2 \\ \cdot \\ \cdot \\ \cdot \\ \dot{x}_n \end{bmatrix} = \begin{bmatrix} a_{11} & a_{12} & \cdots & a_{1n} \\ a_{21} & a_{22} & \cdots & a_{2n} \\ \cdot & & & \\ \cdot & & & \\ \cdot & & & \\ a_{n1} & a_{n2} & \cdots & a_{nn} \end{bmatrix} \begin{bmatrix} x_1 \\ x_2 \\ \cdot \\ \cdot \\ \cdot \\ x_n \end{bmatrix} + \begin{bmatrix} b_{11} & \cdots & b_{1i} \\ b_{21} & \cdots & b_{2i} \\ \cdot & & \\ \cdot & & \\ \cdot & & \\ b_{n1} & \cdots & b_{ni} \end{bmatrix} \begin{bmatrix} r_1 \\ r_2 \\ \cdot \\ \cdot \\ \cdot \\ r_i \end{bmatrix} \tag{9.3.3}
$$

From (9.3.3) we obtain the vector state equation

$$
\dot{\mathbf{x}} = \mathbf{A}\mathbf{x} + \mathbf{B}\mathbf{r} \tag{9.3.4}
$$

where the state vector is given by

$$
\mathbf{x} = \begin{bmatrix} x_1 \\ x_2 \\ \cdot \\ \cdot \\ \cdot \\ x_n \end{bmatrix}
$$

and the input vector is

$$
\mathbf{r} = \begin{bmatrix} r_1 \\ r_2 \\ \cdot \\ \cdot \\ \cdot \\ r_i \end{bmatrix}
$$

where $\mathbf{A}$ is an $n \times n$ matrix and B is an $n \times i$ matrix, both having constant coefficients.

The output vector equation is given by

$$
\begin{bmatrix} y_1 \\ y_2 \\ \cdot \\ \cdot \\ \cdot \\ y_m \end{bmatrix} = \begin{bmatrix} c_{11} & c_{12} & \cdots & c_{1n} \\ \cdot & & & \\ \cdot & & & \\ \cdot & & & \\ c_{m1} & c_{m2} & \cdots & c_{mn} \end{bmatrix} \begin{bmatrix} x_1 \\ \cdot \\ \cdot \\ \cdot \\ x_n \end{bmatrix} \tag{9.3.5}
$$

or

$$
\mathbf{y} = \mathbf{C}\mathbf{x} \tag{9.3.6}
$$

with the output vector as

$$
\mathbf{y} = \begin{bmatrix} y_1 \\ y_2 \\ \cdot \\ \cdot \\ \cdot \\ y_m \end{bmatrix}
$$

and the C matrix is $m \times n$.

From equations (9.3.4) and (9.3.6) we find a very compact form of the general control system. For this reason the state-variable approach is more adaptable to computer solution and control. The nth-order differential equation is now n first-order coupled differential equations. The solution to first-order differential equations is quite simple and is the topic of the next section.

9.4 SOLUTION OF THE STATE-VECTOR EQUATION

If we consider state-vector equation (9.3.4) as one-dimensional, we obtain

$$
\dot{x} = ax + br \tag{9.4.1}
$$

Taking Laplace transforms of equation (9.4.1) with zero initial conditions, we have

$$
sX(s) = aX(s) + bR(s)
$$

and therefore

$$
X(s) = \frac{b}{s - a} R(s) \tag{9.4.2}
$$

The inverse Laplace transform of equation (9.4.2) provides the solution, using the convolution property:

$$
x(t) = \int_0^t e^{a(t-\tau)} br(\tau) \, d\tau \tag{9.4.3}
$$

The solution to the multidimensional vector differential equation would therefore be

$$
\mathbf{x}(t) = \int_0^t e^{\mathbf{A}(t-\tau)} \mathbf{B}\mathbf{r}(\tau) \, d\tau \tag{9.4.4}
$$

where

$$e^{\mathbf{A}t} = \sum_{k=0}^{\infty} \frac{\mathbf{A}t^k}{k!} \tag{9.4.5}$$

represents the matrix exponential function which is convergent for all t and any matrix $\mathbf{A}$.

The solution given by equation (9.4.4) can be attained by Laplace-transforming equation (9.3.4) with zero initial conditions—that is,

$$s\mathbf{X}(s) = \mathbf{A}\mathbf{X}(s) + \mathbf{B}\mathbf{R}(s)$$

and therefore

$$\mathbf{X}(s) = [s\mathbf{I} - \mathbf{A}]^{-1} \cdot \mathbf{B}\mathbf{R}(s) \tag{9.4.6}$$

where I represents the identity matrix.   If we find the inverse Laplace transform, we obtain

$$\mathbf{x}(t) = \int_0^t \mathbf{\Phi}(t - \tau) \cdot \mathbf{B} \cdot r(\tau)\, d\tau \tag{9.4.7}$$

where

$$\mathbf{\Phi}(t) = \mathcal{L}^{-1}\{[s\mathbf{I} - \mathbf{A}]^{-1}\} \tag{9.4.8}$$

and is known as the **state-transition matrix**.   If initial conditions had been considered, we would find the state-transition matrix representing the unforced system response.   If we compare equation (9.4.4) with (9.4.7), we find

$$\mathbf{\Phi}(t) = e^{\mathbf{A}t} \tag{9.4.9}$$

From equation (9.4.9) we find the state-transition matrix available by numerical methods.   If an approximation is required, we can use a finite number of terms and evaluate via computer program.   The reader is encouraged to generate a computer program that will use ten terms in the approximation of the state-transition matrix.   From equation (9.4.8) we can accurately calculate the transition matrix using classical methods.

EXAMPLE 9.3

Consider the following control system.   Find the state-transition matrix, using equation (9.4.8) and the system output, if the input is a unit step function.   All initial conditions are equal to zero.

$$\begin{bmatrix} \dot{x}_1 \\ \dot{x}_2 \end{bmatrix} = \begin{bmatrix} -3 & 1 \\ -2 & 0 \end{bmatrix} \begin{bmatrix} x_1 \\ x_2 \end{bmatrix} + \begin{bmatrix} 4 \\ -5 \end{bmatrix} r(t)$$

$$y(t) = [1 \;-\; 1] \begin{bmatrix} x_1 \\ x_2 \end{bmatrix}$$

If we begin with $[sI - A]$, we have

$$[sI - A] = \begin{bmatrix} s & 0 \\ 0 & s \end{bmatrix} - \begin{bmatrix} -3 & 1 \\ -2 & 0 \end{bmatrix}$$

$$= \begin{bmatrix} s + 3 & -1 \\ 2 & s \end{bmatrix}$$

and now the inverse matrix is

$$[sI - A]^{-1} = \begin{bmatrix} s + 3 & -1 \\ 2 & s \end{bmatrix}^{-1}$$

$$= \begin{bmatrix} \dfrac{s}{s^2 + 3s + 2} & \dfrac{1}{s^2 + 3s + 2} \\ \dfrac{-2}{s^2 + 3s + 2} & \dfrac{s + 3}{s^2 + 3s + 2} \end{bmatrix}$$

and, taking the inverse Laplace transform of each term, we obtain the state-transition matrix:

$$\Phi(t) = \mathcal{L}^{-1}\{[sI - A]^{-1}\} = \begin{bmatrix} (-e^{-t} + 2e^{-2t}) & (e^{-t} - e^{-2t}) \\ (-2e^{-t} + 2e^{-2t}) & (2e^{-t} - e^{-2t}) \end{bmatrix}$$

For comparison it would be interesting to evaluate the calculated state-transition matrix at $t = 1$ sec and also evaluate the exponential matrix using ten terms in the series. From our calculated matrix, we have

$$\Phi(1) = \begin{bmatrix} -0.09721 & 0.2325 \\ -0.4651 & 0.6004 \end{bmatrix}$$

and from ten terms in the series, $t = 1$ sec, we obtain

$$e^A = \begin{bmatrix} -0.09712 & 0.2325 \\ -0.4650 & 0.6004 \end{bmatrix}$$

To find the solution to our system we still have to perform the convolution integrals as indicated by equation (9.4.7). From that equation, our state-transition matrix, and a unit step input $r(t) = 1$ we have

$$\mathbf{x}(t) = \begin{bmatrix} \displaystyle\int_0^\infty (-e^{-(t-\tau)} + 2e^{-2(t-\tau)})\, d\tau & \displaystyle\int_0^\infty (e^{-(t-\tau)}) - e^{-2(t-\tau)})\, d\tau \\ \\ \displaystyle\int_0^\infty (-2e^{-(t-\tau)} + 2e^{-2(t-\tau)})\, d\tau & \displaystyle\int_0^\infty (2e^{-(t-\tau)} - e^{-2(t-\tau)})\, d\tau \end{bmatrix} \begin{bmatrix} 4 \\ \\ -5 \end{bmatrix}$$

Performing the integration and simplifying, we obtain

$$\mathbf{x}(t) = \begin{bmatrix} (9e^{-t} - 6.5e^{-2t} - 2.5) \\ (18e^{-t} - 6.5e^{-2t} - 11.5) \end{bmatrix}$$

with the system output as

$$y(t) = x_1(t) - x_2(t)$$

yielding the simplified system output

$$y(t) = 9(1 - e^{-t})$$

This example has provided a solution to the state-vector equation.   If one is to use the analysis developed in the frequency domain, then a transfer function is required.   Let's investigate the transfer functions in state space.

9.5 TRANSFER FUNCTIONS IN STATE SPACE

In our work with the frequency domain we found the transfer function important in predicting the stability of a system.   The characteristic equation provided the location of the system poles and a prediction of system response.   If we refer to equation (9.3.6) and take Laplace transforms of both sides, we have

$$\mathbf{Y}(s) = \mathbf{C}\mathbf{X}(s) \tag{9.5.1}$$

If we substitute equation (9.4.6) into (9.5.1), we obtain

$$\mathbf{Y}(s) = \mathbf{C} \cdot [s\mathbf{I} - A]^{-1} \cdot \mathbf{BR}(s) \tag{9.5.2}$$

and our transfer function $\mathbf{P}(s)$ is given by

$$\mathbf{P}(s) = \mathbf{C} \cdot [s\mathbf{I} - A]^{-1} \cdot \mathbf{B} \tag{9.5.3}$$

From equation (9.5.3) it is evident that our transfer function is an $m \times i$ matrix with the following entries:

$$\mathbf{P}(s) = \begin{bmatrix} P_{11}(s) & P_{12}(s) & \cdots & P_{1i}(s) \\ \cdot & & & \\ \cdot & & & \\ \cdot & & & \\ P_{m1} & P_{m2}(s) & \cdots & P_{mi}(s) \end{bmatrix} \tag{9.5.4}$$

where

$$P_{mi}(s) = \frac{Y_m(s)}{R_i(s)} \tag{9.5.5}$$

From equation (9.5.3) it is evident that the denominator of the transfer-function matrix will be given by its determinant. Therefore our characteristic equation is

$$\det [s\mathbf{I} - \mathbf{A}] = 0 \tag{9.5.6}$$

From the transfer-function matrix we can predict the system outputs relative to each input. The characteristic equation can predict the system response and overall stability. This process is similar to the analysis carried out in the previous chapters. The reader should appreciate the compactness of the state-variable vector equation and its adaptability to complex control systems.

EXAMPLE 9.4

In Example 9.3 we found the system response using time-domain techniques. Show that the same output is attained by using the frequency-domain and transfer functions. The input is $R(s) = 1/s$.

SOLUTION To find the transfer-function matrix we use Equation (9.5.3). Substituting the information from Example 9.3, we have

$$\mathbf{P}(s) = [1 \quad -1] \begin{bmatrix} \dfrac{s}{s^2 + 3s + 2} & \dfrac{1}{s^2 + 3s + 2} \\ \dfrac{-2}{s^2 + 3s + 2} & \dfrac{s + 3}{s^2 + 3s + 2} \end{bmatrix} \begin{bmatrix} 4 \\ -5 \end{bmatrix}$$

and, simplifying, we obtain

$$P(s) = \frac{9}{s + 1}$$

Using $R(s) = 1/s$, the inverse Laplace transform, we have

$$y(t) = 9(1 - e^{-t})$$

which is the same result as obtained from the time-domain response of Example 9.3.

This example generated one transfer function, and it would be beneficial to investigate a more involved system.

EXAMPLE 9.5

Given the following state equations for a control system, find the characteristic equation and the transfer-function matrix.

$$\begin{bmatrix} \dot{x}_1 \\ \dot{x}_2 \\ \dot{x}_3 \end{bmatrix} = \begin{bmatrix} -4 & 0 & 2 \\ -1 & -1 & 0 \\ 3 & 0 & -3 \end{bmatrix} \begin{bmatrix} x_1 \\ x_2 \\ x_3 \end{bmatrix} + \begin{bmatrix} -3 & -2 \\ 4 & 1 \\ 0 & 0 \end{bmatrix} \begin{bmatrix} r_1 \\ r_2 \end{bmatrix}$$

$$\begin{bmatrix} y_1 \\ y_2 \end{bmatrix} = \begin{bmatrix} 1 & 1 & 1 \\ -1 & 0 & 1 \end{bmatrix} \begin{bmatrix} x_1 \\ x_2 \\ x_3 \end{bmatrix}$$

SOLUTION   The $[s\mathbf{I} - \mathbf{A}]$ matrix is

$$[s\mathbf{I} - \mathbf{A}] = \begin{bmatrix} s+4 & 0 & -2 \\ 1 & s+1 & 0 \\ -3 & 0 & s+3 \end{bmatrix}$$

with the inverse being

$$[s\mathbf{I} - \mathbf{A}]^{-1} = \begin{bmatrix} \dfrac{s+3}{s^2+7s+6} & 0 & \dfrac{2}{s^2+7s+6} \\ \dfrac{-(s+3)}{(s+1)(s^2+7s+6)} & \dfrac{1}{s+1} & \dfrac{-2}{(s+1)(s^2+7s+6)} \\ \dfrac{3}{s^2+7s+6} & 0 & \dfrac{s+4}{s^2+7s+6} \end{bmatrix}$$

Using equation (9.5.3), we have

$$\mathbf{P}(s) =$$

$$\begin{bmatrix} 1 & 1 & 1 \\ -1 & 0 & 1 \end{bmatrix} \begin{bmatrix} \dfrac{s+3}{s^2+7s+6} & 0 & \dfrac{2}{s^2+7s+6} \\ \dfrac{-(s+3)}{(s+1)(s^2+7s+6)} & \dfrac{1}{s+1} & \dfrac{-2}{(s+1)(s^2+7s+6)} \\ \dfrac{3}{s^2+7s+6} & 0 & \dfrac{s+4}{s^2+7s+6} \end{bmatrix} \begin{bmatrix} -3 & -2 \\ 4 & 1 \\ 0 & 0 \end{bmatrix}$$

which, upon simplification, becomes

$$\mathbf{P}(s) = \begin{bmatrix} \dfrac{s^2+10s+15}{(s+1)(s^2+7s+6)} & \dfrac{-s(s+5)}{(s+1)(s^2+7s+6)} \\ \dfrac{3s}{s^2+7s+6} & \dfrac{2s}{s^2+7s+6} \end{bmatrix}$$

The characteristic equation is given by the determinant of the matrix ($[s\mathbf{I} - \mathbf{A}]$) and set equal to zero. That is,

$$(s+1)(s^2+7s+6) = 0$$

From the characteristic equation we can predict the stability and time response of the system. In terms of the state variables it would be nice to know if we have direct input to each state variable and if we can measure the output of each state variable. These properties are known as *controllability* and *observability*, respectively, and are the topic of the next section.

9.6 CONTROLLABILITY AND OBSERVABILITY

The ideas of controllability and observability are important in the design of control systems using the state-variable approach. In terms of controllability, we have

> *A system is totally **controllable** if every state can be driven to a required output with a control input.*

The concept of controllability is not obvious from the state equations and requires diagonalization of the **A** matrix. If we diagonalize the **A** matrix, we obtain a decoupled system, and the controllability becomes evident. Figure 9.7 shows a decoupled system which is totally uncontrollable. It is evident that state variable $x_1(t)$ has no input. This implies we have no control of its output, and this mode is considered uncontrollable. Each decoupled equation is considered to be a **mode**. We will not investigate the diagonalization of the **A** matrix, as a mathematical test is available.

In terms of observability, we have

> *A system is totally **observable** if every state output can be measured at any of the system outputs.*

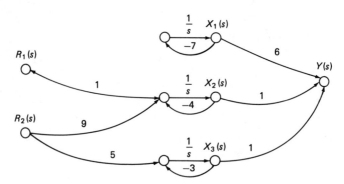

Figure 9.7   Decoupled system that is totally uncontrollable

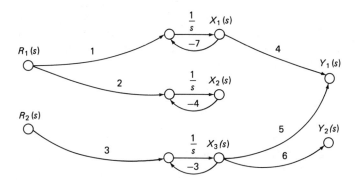

Figure 9.8   Decoupled system that is totally unobservable

The observability is evident from the decoupled state equations, and Fig. 9.8 portrays an unobservable mode. From Fig. 9.8 we find that the output of state variable $x_2(t)$ is not measurable at any of the system outputs. This implies an unobservable mode and therefore an unobservable system. The reader should understand that the performance of a state variable is not evident if it cannot be measured. We will investigate a mathematical test for observability.

If we consider our vector-state equation given by equation (9.3.4) and generate the matrix $\mathbf{M}_c$,

$$\mathbf{M}_c = [\mathbf{B} \,|\, \mathbf{AB} \,|\, \ldots \,|\, \mathbf{A}^{n-1}\mathbf{B}] \tag{9.6.1}$$

the control system is totally controllable if its controllability matrix $\mathbf{M}_c$ is of full rank.

EXAMPLE 9.6

Consider the following system. Use the controllability matrix to investigate the controllability of the system.

$$\begin{bmatrix} \dot{x}_1 \\ \dot{x}_2 \\ \dot{x}_3 \end{bmatrix} = \begin{bmatrix} -2 & 1 & 2 \\ 4 & 0 & 3 \\ 1 & -1 & 0 \end{bmatrix} \begin{bmatrix} x_1 \\ x_2 \\ x_3 \end{bmatrix} + \begin{bmatrix} 0 & 4 \\ -5 & 0 \\ 0 & 0 \end{bmatrix} \begin{bmatrix} r_1 \\ r_2 \end{bmatrix}$$

SOLUTION   For this system the controllability matrix becomes

$$\mathbf{M}_c = [\mathbf{B} \,|\, \mathbf{AB} \,|\, \ldots \,|\, \mathbf{A}^2\mathbf{B}]$$

and after multiplication we have

$$M_c = \begin{bmatrix} 0 & 4 & -5 & -8 & 20 & 40 \\ -5 & 0 & 0 & 16 & -5 & -20 \\ 0 & 0 & 5 & 4 & -5 & -24 \end{bmatrix}$$

To be of full rank the controllability matrix must have at least one $n \times n$ matrix with a nonzero determinant. If we consider the following 3×3 matrix, we find the determinant nonzero; therefore the system is totally controllable.

$$
\begin{bmatrix}
0 & 4 & -5 \\
-5 & 0 & 0 \\
0 & 0 & 5
\end{bmatrix}
$$

If we consider equations (9.3.4) and (9.3.6) and generate the matrix $\mathbf{M}_o$,

$$
\mathbf{M}_o =
\begin{bmatrix}
\mathbf{C} \\
\hline
\mathbf{CA} \\
\hline
\cdot \\
\cdot \\
\cdot \\
\mathbf{CA}^{n-1}
\end{bmatrix}
\tag{9.6.2}
$$

the control system is totally observable if its observability matrix ($\mathbf{M}_o$) is of full rank.

EXAMPLE 9.7

Consider the following state equations. Use the observability matrix to investigate the observability of the system.

$$
\begin{bmatrix}
\dot{x}_1 \\
x_2 \\
x_3
\end{bmatrix}
=
\begin{bmatrix}
2 & 1 & 0 \\
-3 & 0 & 1 \\
4 & 0 & 0
\end{bmatrix}
\begin{bmatrix}
x_1 \\
x_2 \\
x_3
\end{bmatrix}
+
\begin{bmatrix}
1 \\
1 \\
1
\end{bmatrix}
r(t)
$$

$$
y(t) = \begin{bmatrix} 0 & 0 & 1 \end{bmatrix}
\begin{bmatrix}
x_1 \\
x_2 \\
x_3
\end{bmatrix}
$$

SOLUTION   For this system the observability matrix becomes

$$
\mathbf{M}_o =
\begin{bmatrix}
\mathbf{C} \\
\mathbf{CA} \\
\mathbf{CA}^2
\end{bmatrix}
$$

and after multiplication we have

$$\mathbf{M}_o = \begin{bmatrix} 0 & 0 & 1 \\ 4 & 0 & 0 \\ 8 & 4 & 0 \end{bmatrix}$$

Since the determinant of the observability matrix is nonzero, the matrix is of full rank and the system is totally observable.

State-variable analysis and design is a very complicated and modern approach. The information presented here is intended to introduce the reader to this fascinating area of modern control theory. The mathematical content and proofs have been kept to a minimum to allow easy reading.

This concludes our study of continuous control systems. The reader is advised to review, if necessary, in order to get a firm grasp of the basic principles of continuous control systems, as they will become instrumental in the understanding of discrete control systems.

9.7 SUMMARY

In this chapter we introduced the state-variable approach to control systems. It was evident that large systems with multiple inputs and outputs could be represented in compact form in this analysis. The matrix approach makes computer computation easier and more efficient. The design and analysis of nonlinear and time-varying control systems is possible.

The phase-variable diagram was introduced as a graphical way of transforming our transfer function into state variables. The state equations were generated and found to be coupled first-order differential equations. The time solution to these equations was made possible with the exponential matrix and classical transition matrix. The first provided a numerical solution, the second a mathematical one.

To apply our classical frequency-analysis methods we generated the transfer-function matrix. The determinant of the $[s\mathbf{I} - \mathbf{A}]$ matrix became the characteristic equation for the system. Now we could predict system response and stability. The ideas of controllability and observability were introduced. If a system is to be totally controllable, then all the states require an input. If this is not possible, then the state will become uncontrollable. If a system is to be totally observable, then all the state outputs should be measurable at any of the outputs. The ideas of controllability and observability were evident in a decoupled system and required mathematical testing in coupled systems. The controllability and observability matrixes were defined for this purpose.

PROBLEMS

9.1. Consider the simulation diagram shown in Fig. P9.1. If we indicate the output of each integrator as a state variable, taking into consideration the inversion, we find a resemblance to the phase-variable simulation diagram.

(a) Generate the equivalent phase-variable simulation diagram.

(b) Obtain the set of first-order differential equations in terms of the state variables.

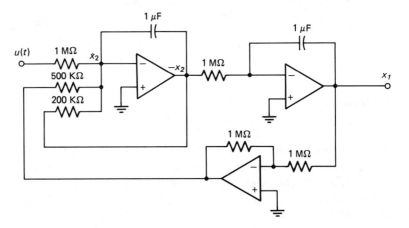

Figure P9.1   Simulation diagram for Problem 9.1

9.2. The idea of state variables is that if we know the present state of a system, then we can predict the future of that state with a given input. Consider the physical system in Fig. 6.9. The future position of the mass can be obtained if we know the present position (or state x_1) along with the velocity (or state x_2) of the mass. Therefore this system requires two state variables to fully describe it. This implies

$$x_1 = y \Rightarrow \dot{x}_1 = x_2$$

$$x_2 = \dot{y} \Rightarrow \dot{x}_2 = \ddot{y}$$

(a) Use the definition of the state variables to generate two first-order differential equations from equation (6.5.7).

(b) From the transfer function given by equation (6.5.9), provide the phase-variable simulation diagram. Provide the state equations from the diagram. How do they compare to the equations generated in part (a)?

9.3. In passive circuits the number of state variables required to completely describe the system is equal to the number of independent energy-storage components. Consider the *RLC* circuit given in Fig. P9.3.

    (a) If we take the voltage across the capacitor as state x_1 and the current in the circuit as state x_2, find the two first-order differential equations.

    (b) Now use Laplace transforms to find the system transfer function from Fig. P9.3, the ratio of the capacitor voltage to input voltage. Generate the phase-simulation diagram and the state equations. How do these compare with the equations in part (a)?

    (c) Use the equations from part (a) to find the state-transfer function. How does it compare to the transfer function generated in part (b)? Is there more than one way to pick the state variables for the system?

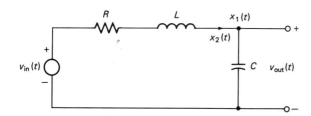

Figure P9.3 *RLC* circuit for Problem 9.3

9.4. Consider the circuit in Fig. P9.4. If state variable x_1 is the voltage across the capacitor and x_2 is the current through the inductor, show that the **A** and **B** matrices are

$$\mathbf{A} = \begin{bmatrix} -\dfrac{1}{RC} & 0 \\ 0 & -\dfrac{R}{L} \end{bmatrix}, \qquad \mathbf{B} = \begin{bmatrix} \dfrac{1}{2RC} & \dfrac{1}{2RC} \\ \dfrac{1}{2L} & -\dfrac{1}{2L} \end{bmatrix}$$

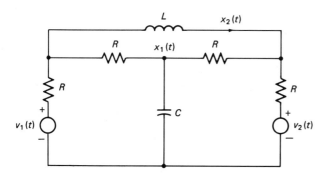

Figure P9.4 Circuit for Problem 9.4

9.5. Given the following state-vector equations, what is the system damping ratio?

$$\begin{bmatrix} \dot{x}_1 \\ \dot{x}_2 \end{bmatrix} = \begin{bmatrix} -6 & 4 \\ -4 & 2 \end{bmatrix} \begin{bmatrix} x_1 \\ x_2 \end{bmatrix} + \begin{bmatrix} 4 \\ -7 \end{bmatrix} r(t)$$

$$y = \begin{bmatrix} 1 & -1 \end{bmatrix} \begin{bmatrix} x_1 \\ x_2 \end{bmatrix}$$

9.6. Given the following state-vector equations, find the state transfer-function matrix.

$$\begin{bmatrix} \dot{x}_1 \\ \dot{x}_2 \end{bmatrix} = \begin{bmatrix} -2 & 3 \\ -1 & -1 \end{bmatrix} \begin{bmatrix} x_1 \\ x_2 \end{bmatrix} + \begin{bmatrix} 4 & 0 \\ -5 & 6 \end{bmatrix} \begin{bmatrix} r_1 \\ r_2 \end{bmatrix}$$

$$y = \begin{bmatrix} 7 & 8 \end{bmatrix} \begin{bmatrix} x_1 \\ x_2 \end{bmatrix}$$

9.7. Given the following state-vector equations, find the state transfer-function matrix.

$$\begin{bmatrix} \dot{x}_1 \\ \dot{x}_2 \\ x_3 \end{bmatrix} = \begin{bmatrix} -2 & 0 & 1 \\ -1 & -2 & 0 \\ 3 & 1 & -4 \end{bmatrix} \begin{bmatrix} x_1 \\ x_2 \\ x_3 \end{bmatrix} + \begin{bmatrix} -2 & 3 \\ 5 & 2 \\ 1 & 0 \end{bmatrix} \begin{bmatrix} r_1 \\ r_2 \end{bmatrix}$$

$$\begin{bmatrix} y_1 \\ y_2 \\ y_3 \end{bmatrix} = \begin{bmatrix} 1 & 0 & 1 \\ 2 & -1 & 2 \\ 1 & -1 & 3 \end{bmatrix} \begin{bmatrix} x_1 \\ x_2 \\ x_3 \end{bmatrix}$$

9.8. Can a system be totally controllable and not totally observable?

9.9. Can a system be totally observable and not totally controllable?

9.10. Can a system be totally controllable, observable, and unstable?

9.11. Consider the following state-vector equations:

$$\begin{bmatrix} \dot{x}_1 \\ \dot{x}_2 \end{bmatrix} = \begin{bmatrix} 2 & 3 \\ 6 & -1 \end{bmatrix} \begin{bmatrix} x_1 \\ x_2 \end{bmatrix} + \begin{bmatrix} 1 \\ -2 \end{bmatrix} r(t)$$

$$y = \begin{bmatrix} 1 & 0 \end{bmatrix} \begin{bmatrix} x_1 \\ x_2 \end{bmatrix}$$

(a) Is the system totally controllable?
(b) Is the system totally observable?

9.12 Consider the following state-vector equations:

$$\begin{bmatrix} \dot{x}_1 \\ \dot{x}_2 \end{bmatrix} = \begin{bmatrix} 0 & 1 \\ -1 & 1 \end{bmatrix} \begin{bmatrix} x_1 \\ x_2 \end{bmatrix} + \begin{bmatrix} 0 \\ 1 \end{bmatrix} r(t)$$

$$y = \begin{bmatrix} 1 & 0 \end{bmatrix} \begin{bmatrix} x_1 \\ x_2 \end{bmatrix}$$

(a) Is the system totally controllable?
(b) Is the system totally observable?
(c) Is the system stable?
(d) If this represents a unity-feedback system, provide the open-loop transfer function.

9.13. With reference to the definition of a state given in Problem 9.2, what is the importance of controllability and observability?

9.14. Consider the following state-vector equations:

$$\begin{bmatrix} \dot{x}_1 \\ \dot{x}_2 \\ \dot{x}_3 \end{bmatrix} = \begin{bmatrix} 0 & 1 & 0 \\ 0 & 0 & 1 \\ -4 & -7 & -4 \end{bmatrix} \begin{bmatrix} x_1 \\ x_2 \\ x_3 \end{bmatrix} + \begin{bmatrix} 0 \\ 0 \\ 1 \end{bmatrix} r(t)$$

$$y = \begin{bmatrix} 2 & 0 & 0 \end{bmatrix} \begin{bmatrix} x_1 \\ x_2 \\ x_3 \end{bmatrix}$$

(a) Provide a phase-variable simulation diagram.
(b) Find the system transfer function.
(c) Find and graph the system response due to a unit step input.
(d) Comment on observability and controllability.
(e) Predict the time response of the remaining two state variables.
(f) If part (b) represents an open-loop transfer function, what are the gain and phase margin?
(g) Provide a root locus for the transfer function used in part (f).

9.15. Provide an analog simulation for the system in Problem 9.14. Compare the outputs of the state variables with those predicted in Problem 9.14.

9.16. Generate a computer program (language of your choice) to evaluate the first ten terms of the exponential matrix to yield the state-transition matrix of Problem 9.14. Use $t = 1$ second to see if your program is operational. If twenty terms are used, what is the error?

Part II
Discrete

10

Digital Control Systems

10.1 INTRODUCTION

It would appear that the digital implementation of control systems is a new technique brought about by the inception of the microprocessor. One would have to agree on the impact of the microprocessor, yet question the uniqueness of the technique. What we have here is a technique known as the approximation of a continuous system with a finite number of points. It is apparent that the approximation will emulate the original system as the number of points approaches infinity. The latter statement is very familiar in the development of the definition of the integral and derivative in the field of calculus. This implies that all continuous systems began as a finite set of points and through the idea of limits, enabled the mathematician to represent the same in closed form. Now we find that all systems are digital in nature and only through mathematical means mature to the continuous form.

So why do we consider digital control as the modern control scheme? The answer is quite simple. In the past the tools available to mankind forced the representation of systems in a closed or continuous form. As technology matured and the microprocessor was developed, we found that the tools available could in fact reproduce the continuous system via approximation. It is more of a realization of an idea whose time has come as opposed to the introduction of a new idea.

This chapter will elaborate on this technique and introduce the tools which made it possible—namely, the sample-and-hold (S/H) circuit, the analog-to-digital converter (A/D), the microprocessor, and the digital-to-analog converter (D/A).

10.2 APPROXIMATION OF A CONTINUOUS SYSTEM

Any compensation network for a control system can be shown as in Fig. 10.1, where

$$e(t) \quad \rightarrow \quad \text{error at time } t$$

$$u(t) \quad \rightarrow \quad \text{control action at time } t$$

$$N \leq K$$

$$A_*, B_* \quad \text{are constants}$$

from which the following differential equation is generated:

$$B_0 u(t) = \frac{d^N}{dt^N} e(t) + A_{N-1} \frac{d^{N-1}}{dt^{N-1}} e(t) + \cdots + A_0 e(t)$$

$$- \frac{d^K}{dt^K} u(t) - B_{K-1} \frac{d^{K-1}}{dt^{K-1}} u(t) - \cdots - B_1 \frac{d}{dt} u(t) \qquad (10.2.1)$$

The real-time solution of (10.2.1) represents the control action required in our control system at time t. Let us define a function $f(t)$ such that f_k represents the point at which the derivative is to be calculated and similarly f_{k-N} represents N points equidistant by T time units from our present point f_k. Figure 10.2 (next page) portrays the function $f(t)$.

To find the derivative at f_k we have by definition:

$$\frac{d}{dt} f_k \overset{\triangle}{=} \lim_{T \to 0} \frac{f_k - f_{k-1}}{T} \qquad (10.2.2)$$

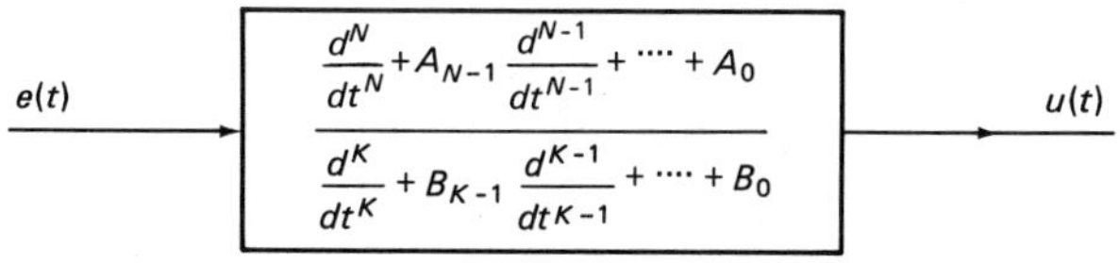

Figure 10.1 Typical compensation network

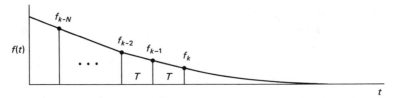

Figure 10.2 Arbitrary function $f(t)$

and the second derivative is given by

$$\frac{d^2}{dt^2} f_k \triangleq \lim_{T^2 \to 0} \frac{f_k - 2f_{k-1} + f_{k-2}}{T^2} \qquad (10.2.3)$$

yielding a general form for the Nth derivative:

$$\frac{d^N}{dt^N} f_k \triangleq \lim_{T^N \to 0} \frac{\sum_{n=1}^{N+1} (-1)^{n+1} P_\triangle(N, n) f_{k-n+1}}{T^N} \qquad (10.2.4)$$

where $f_\triangle(N, n)$ represents the Nth row and nth element of Pascal's triangle. If we redefine our function $f(t)$, in terminology only, we have the present sample f_k with the Nth past sample being f_{k-N}. The time increment between samples is given by the sampling time T. Using finite time T and the proper subscripts, we can rewrite equation (10.2.1) with successive substitutions of equation (10.2.4) to yield

$$u_k = C_N e_{k-N} + C_{N-1} e_{k-N+1} + \cdots C_0 e_k - D_K u_{k-K} - \cdots - D_1 u_{k-1} \qquad (10.2.5)$$

where C_*, D_* are the appropriate constants. On close inspection of (10.1.5) it is evident that we have generated a difference equation, which, if T approaches zero, will generate the solution of the differential equation given by (10.2.1). This result is quite interesting, in that we have developed an iterative method of solving the typical compensation network required in control systems. As computers lend themselves to iteration procedures, we have found a method of incorporating the computer in the closed-loop control system. Simply put, the computer can calculate the present control action u_k based on N past samples of the error and K past samples of the control action with the appropriate sampling time to insure stability. An example will clarify this approximation of a differential equation via a difference equation.

EXAMPLE 10.1.

Consider the compensation network given in Fig. 10.3 both in time and in the frequency domain with initial conditions at zero.

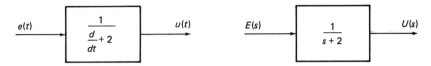

Figure 10.3 Compensation network example

SOLUTION Using (10.2.1), we generate the compensation differential equation

$$2u(t) = e(t) - \frac{d}{dt} u(t) \qquad (10.2.6)$$

yielding a time solution of

$$u(t) = \frac{1}{2} (1 - e^{-2t}) \qquad (10.2.7)$$

subject to a step input ($e(t) = 1$, $E(S) = 1/s$). To find the appropriate difference equation we use equation (10.2.4) to approximate the derivatives in equation (10.2.6). This generates the following control action u_k:

$$u_k = \frac{1}{1 + 2T} (Te_k + u_{k-1}) \qquad (10.2.8)$$

To find the solution to 10.2.8 we can use the BASIC program given in Fig. 10.4 with a sampling time of 0.1 second.

```
  5    REM  *** BASIC PROGRAM TO APPROXIMATE CONTROL ACTION ***
 10    NSAMPLES  = 30
 20    T         = .1
 25    REM *** STEP INPUT ***
 28    ERR = 1
 30    FOR  I = 1 TO NSAMPLES
 40         REM   *** SAVE PREVIOUS SAMPLE ***
 50         U(1) = U(0)
 60         REM   *** GENERATE PRESENT CONTROL ACTION ***
 70         U(0) = 1/(1 + 2*T)*(T*ERR + U(1))
 80         PRINT "TIME = ";T*I,"CONTROL ACTION = ";U(0)
 90    NEXT I
100    END
```

Figure 10.4 BASIC program to approximate control action

A quick look at the graph in Fig. 10.5 provides definite proof that our computer approximation is very close to the original continuous solution.

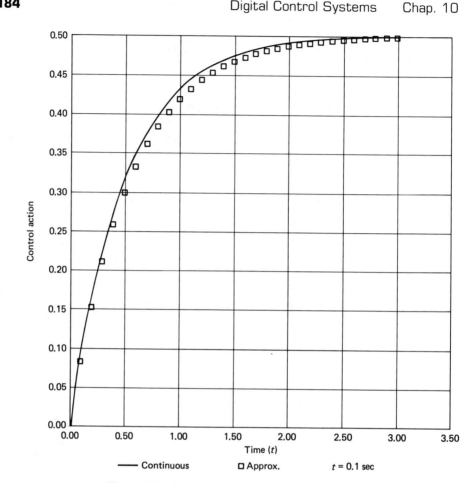

Figure 10.5   Graph of control action for Example 10.1

It is left as an exercise for the reader to run the program in Fig. 10.4 with different sampling times and to compare the results with the continuous response. The reader should also be wondering about the implications of this approximation if generated by a finite mathematical element such as the N-bit microprocessor.

The preceding example enabled us to appreciate that a computer can approximate the solution to a differential equation by using a difference equation. With this in place, it is important to understand the hardware elements required to make this theory practical. We now turn our attention to these elements.

10.3 HARDWARE ELEMENTS OF A DIGITAL CONTROL SYSTEM

Figure 10.6 provides a digital implementation of our compensation network as given in Fig. 10.1.

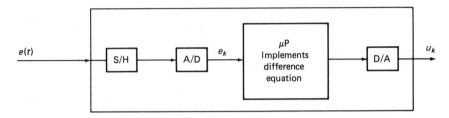

Figure 10.6   Digital version of compensation network

The sample-and-hold block (S/H) provides a practical impulse sample to guarantee an accurate sample of high frequencies. This holding technique provides a *zero-order hold process*, a term which will become important in later chapters, so that the analog-to-digital (A/D) converter can digitize the sample. Once the sample is digitized, the microprocessor (μP) can perform the iterative difference equation and generate the present control action. In order to accommodate the continuous nature of practical systems, the control action is converted to an analog signal via the digital-to-analog (D/A) converter.

This represents a brief overview of the hardware components involved. Now we will investigate each component in detail. Let's begin with the A/D converter.

A/D Conversion

The most popular method of conversion is via successive approximation. If the reader is unfamiliar with this method, a revisit to a digital text is recommended. We are concerned here only with the limitations and practical applications of the device itself.

Let's look at some common terms associated with A/D converters. We will assume a converter which responds to $\pm V$ volts and provides N bits of conversion.

Resolution (RES).   *Smallest detectable change at the output. In* *our case this is given by*

$$\text{RES} = \frac{V}{2^{N-1}} \tag{10.3.1}$$

Conversion time (T_c). *Sometimes known as* **aperture time** *(T_a). This represents the time required to perform the conversion. This quantity is provided by the manufacturer and is indicated in hardware with an end-of-conversion (EOC) pulse. [Note: Input should not change by more than one-half the least significant bit (LSB) during this time or erroneous conversions will result.]*

Accuracy. *This is affected by zero drift, power-supply sensitivity, gain error, nonlinearity error, and noise. In most control systems the accuracy will be input-signal related and not device related.*

To appreciate these terms, let's take a look at a typical A/D converter.

EXAMPLE 10.3

Given the following successive-approximation A/D converter specifications:

$$T_c = 10 \ \mu s$$

$$V = \pm 5 \text{ volts}$$

$$N = 16 \text{ bits}$$

Calculate:

1. Resolution of converter.
2. Bandwidth of input to guarantee accurate conversion of sample.

SOLUTION

1. Using equation (10.3.1), we have

$$\text{RES} = \frac{5}{2^{16-1}} = 152.6 \ \mu V \tag{10.3.2}$$

2. This is an interesting question, and its answer will explain the need for the S/H block. As stated before, the input can only change by one-half the LSB during the conversion time. Let's define the input signal $v(t) = V \sin \omega t$ and indicate the change in the signal during conversion as $\Delta v(t)$. Refer to Fig. 10.7.

   Now the rate of change of $v(t)$ is given by

$$\frac{d}{dt} v(t) = V\omega\cos \omega t \tag{10.3.3}$$

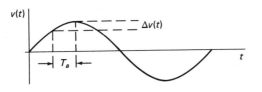

Figure 10.7 Sinusoidal input signal

implying a maximum when cos $\omega t = 1$ or

$$\frac{d}{dt} v(t) = V\omega \qquad (10.3.4)$$

which can be approximated by

$$\frac{\Delta vt}{T_a} = V\omega \qquad (10.3.5)$$

If the input change is to be less than half the LSB, we have

$$V\omega \geq \frac{76.3 \ \mu V}{10 \ \mu s} \qquad (10.3.6)$$

and, solving for frequency ($\omega = 2\pi f$, $V = 5 \ V$), we obtain

$$f \leq 0.24 \ \text{Hz} \qquad (10.3.7)$$

This implies that the input signal bandwidth has to remain below 0.24 hertz to facilitate an accurate conversion.

The converter in the preceding example question can perform 100,000 samples per second and if used alone has a very limited bandwidth. The only way to increase the bandwidth is to effectively decrease the aperture time of the converter, which implies purchase of a quicker converter. Let's assume we have the most cost-effective converter in terms of speed-to-cost ratio and have to find another means of improving the speed of conversion. If we could take a quick snapshot of the input signal and hold it during the aperture time, we would effectively decrease the aperture time and increase the bandwidth. The device in question is a sample-and-hold circuit and will be covered in the next section. The reader is asked if the samples per second will also increase with the introduction of the S/H block.

S/H Block

A typical S/H circuit is given in Fig. 10.8.

With reference to Fig. 10.8 we see the S/H circuit providing an analog memory function via capacitor C. The capacitor charges to $v(t)$ while the

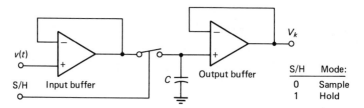

Figure 10.8 S/H circuit

S/H line is at a logical low ("0") and holds the value when the line is at a logical high ("1"). The length of time that the S/H line remains low is considered to be the aperture time (T_a) of the circuit. A typical range for the aperture time can be from 50 nanoseconds to 200 picoseconds.

Let's investigate the results of digitizing the output of the S/H circuit V_k. Figure 10.9 is an enlarged version of Fig. 10.7 to reflect the sample-and hold-function.

From Fig. 10.9 we find the input signal $v(t)$ changing for a time equivalent to the aperture time of the S/H circuit only. If we consider our analog-to-digital converter, specified in Example 10.3, having a sample-and-hold circuit with an aperture time of 250 picoseconds, we can use equation (10.3.6) to calculate the new bandwidth. This yields

$$f = \frac{76.3 \ \mu V}{2\pi \cdot 5 \times 250 \times 10^{-12}}$$

$$= 9.7 \ \text{KHz} \tag{10.3.8}$$

This implies that we can sample a 9.7-KHz sinewave accurately, which is more realistic as compared to 0.24 Hz. The increased bandwidth is the result of the S/H front end. What about the number of samples per second? If we look at Fig. 10.9, it is obvious that the time of conversion is still given by the aperture time of the analog-to-digital converter. Therefore the number of samples per second remains at 100,000. If your prediction was wrong, it would be worthwhile to study Fig. 10.9 and convince yourself of the fact.

We now have the hardware to take an accurate sample and digitize the

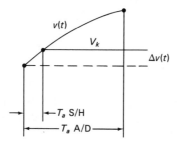

Figure 10.9 Digitizing output of S/H circuit

same. Now we need some intelligence to solve the difference equation to implement our compensation network. Let's introduce the microprocessor.

The Microprocessor

Figure 10.10 provides a block diagram for a typical microprocessor-based control system. The components are interconnected at board level to create a computer control system. It is common to interconnect some of these components at chip level to create a dedicated control microprocessor, or a **microcontroller**. This is a high-speed control processor offering small packaging and high reliability.

Let's define the components given in Fig. 10.10.

Data bus. *This provides a path for information flow between the microprocessor and component. The data bus is N bits wide and is an indication of the accuracy of the microprocessor.*

Address bus. *The microprocessor uses the address bus to select the component before an information exchange occurs. The address bus is 2N bits wide.*

Timing and control. *From a reference system clock, the timing and control section of the microprocessor synchronizes all internal and external data manipulation.*

ALU. *The **arithmetic logic unit** provides the mathematical functions for the microprocessor. In most cases this is limited to addition, subtraction, shifting, rotating, and logical operations.*

Math coprocessor. *If the ALU is insufficient, a math processor*

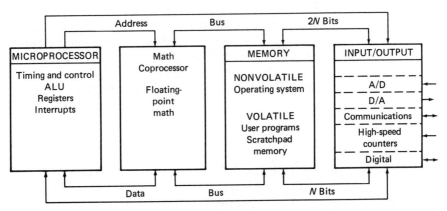

Figure 10.10   Components for microprocessor-based control system

is connected externally to provide high-speed floating-point mathematical functions.

Registers. *These are internal memory locations of the micro-processor. They are dedicated to address manipulation and intermediate data storage.*

Interrupts. *These provide program execution on demand. They can be time or event driven. Each interrupt would invoke a different program or task for execution.*

Nonvolatile memory. *This memory provides information storage with power loss. This area provides a residence for the operating system.*

Operating system. *Creates an interface between the user and microprocessor. The operating system interprets the user's commands, which are high level, and initiates the microprocessor to execute them. In some systems the operating system allows multiple programs or tasks to be resident; this is called* **multitasking**. *These tasks are· executed at specified times or events. If the time is standard clock time (real time), then the operating system is considered to be a* **real-time operating system (RTOS)**.

Volatile memory. *This memory is high-speed and requires battery backup. This area of memory contains the user's programs and any intermediate control variables.*

Input/output (I/O). *External analog signals can be digitized via the A/D converter and acted upon by the microprocessor. The micro-processor can return the analog signals after digital processing by using a D/A converter. The microprocessor can communicate with other com-puter systems or become an element of a network of microprocessor-based control systems. This provides a distributed form of control. The digital I/O provides a way for the microprocessor to implement relay-type logic which exists in the controlled environment. The digital outputs also provide an alternative to analog outputs by using* **pulse-width mod-ulation (PWM)**. *The duty cycle is changed and interpreted by the con-trolled plant as a control action change. The high-speed counters pro-vide digitization of pulse-producing transducers.*

D/A Conversion

This is the final element in our digital version of a compensating network. The D/A converter converts N bits of data to a corresponding voltage. It is always matched with the A/D converter in terms of information and voltage ranges.

The conversion is direct and therefore very fast. The most popular converters use a current weighting scheme such that each bit represents one-half of the current of the previous bit. The reader is encouraged to refer to the manufacturer's data specifications for the specifics of a particular A/D converter.

This completes our brief coverage of the hardware elements of a microprocessor-based control system—to be specific, the digital implementation of a compensating network. As modern technology is constantly producing improved hardware, our discussion has been manufacturer independent. The fundamental principles will remain the same, and an understanding of them will provide the reader with a timeless appreciation for digital control.

In the next chapter we turn our attention to producing computer algorithms for the solution of the compensating network. As in the continuous domain, we require some mathematical tool. That tool is the *z*-transform.

10.4 SUMMARY

This chapter provided a discrete form of the continuous compensation network. The differential equation became a difference equation and promoted iterative solutions. The solutions hinged on samples and storage of previous samples. The idea of sampling indicated an A/D converter as the hardware element required. To provide the storage and manipulation, a microprocessor was required. To return the digital signal to analog form, a D/A was required.

The A/D converter was found to be very bandlimited without the aid of an S/H block. This provided a practical impulse response or sampling function. Figure 10.10 showed the various components involved in a microprocessor-based control system. The reader should be aware that not all components are evident in every digital control system. The components are decided upon by the systems designer at board level, and by the manufacturer at chip level. The result of the manufacturer's endeavors is known as a microcontroller.

Problems

10.1. Refer to Example 10.1.
   (a) Run the program given in Fig. 10.4 with all variables defined as integer values. What happens? Why?
   (b) In part (a) we found the result of integer math. The microprocessor performs integer math and therefore would have this problem. A prac-

tical solution involves scaling.   Consider the following program, a modification of the one in Fig. 10.4.

```
10   CLS
20   N% = 20
30   INPUT "SAMPLING TIME = ",T
40   INPUT "ERROR FULL SCALE = ",E%
50   FOR I% = 1 TO N%
60       U%(1) = U%(0)
70       U%(0) = 10*(E%*T + U%(1))/(10 + 20*T)
80       PRINT U%(0),"TIME = ",T*I%
90   NEXT I%
100  END
```

    (c) Run the program for $T = 0.1$ second and the error full-scale varying
    from 16 to 4096.   This represents a 4-bit to 12-bit representation of
    the error.   Compare to the graph in Fig. 10.5.   Remember that the
    output from the program is relative to the full-scale error.

    (d) Implement the program above on an 8-bit microprocessor of your
    choice.   How does it compare?

10.2. Generate a program to approximate the following control action.   Use a
high-level language and a sampling time of 0.1 second.   Compare to the
calculated time response.   What happens if the sampling time increases?

$$\frac{U(s)}{E(s)} = \frac{1}{(s^2 + 0.2s + 1)}$$

10.3. A control system is using the upper 12 bits of a 16-bit A/D converter.   This
is common practice in industry as a means of eliminating noise.   If the
A/D converter accepts a 10-volt input, what is the resolution of the conversion?   A tach generator provides a 5-volt output at 6000 RPM.   Can
this control system provide 0.1% speed control?

10.4. If the time of conversion T_c for the A/D converter in Problem 10.3 is 54
μs, what is the bandwidth of the converter?

10.5. The control system in Problem 10.3 is using an S/H block with an aperture
time of 500 picoseconds.   What is the bandwidth of the converter?

10.6. A data acquisition system is using a multiplexed scheme.   The system uses
a microprocessor, S/H, A/D converter, and associated logic to acquire data
from N analog channels.   The A/D converter accepts a ± 10-volt input and
is 12-bit.   The microprocessor selects the analog channel, initiates the S/H
block, starts the conversion, and waits for the end-of-conversion pulse.
When the end-of-conversion pulse is present, the microprocessor stores the
value and continues the process on the next channel.   The conversion times
for the A/D converter and S/H block are 8 μs and 1 ns, respectively.   The

microprocessor has a 5-μs overhead for every channel to perform its function. What is the maximum number of analog channels (*N*) possible? What is the maximum bandwidth for any analog channel? What is the sampling frequency of each analog channel?

The bandwidth that you have calculated represents the maximum frequency for the analog signal such that an accurate conversion will result. To recreate any of our analog signals we have to sample at least twice the highest frequency content. This represents the sampling theorem. In our system, what is the highest frequency that can appear at the analog input such that it can be reconstructed? How does this compare with the bandwidth we previously calculated? If we wanted to reconstruct a signal with this bandwidth, what reduction in analog channels (*N*) would be required?

10.7. Investigate the various microcontrollers available on the market. How do they compare to Fig. 10.10, in terms of components contained?

10.8. What is meant by floating-point mathematics?

10.9. Investigate examples of nonvolatile memories. Investigate examples of volatile memories.

10.10. Give an example of an operating system. What are the commands available to the user? How friendly is it?

10.11. Give an example of a time-driven process; an event-driven process.

10.12. A task is basically a program. In a multitasking system we have multiple programs resident. With in-line processing only one task can be functional at any given time. This implies the operating system is multiplexing the tasks. If we have an RTOS, then each task is executed at a specific time. Compare this to Problem 10.6. If each task is performing a control action in a controlled process, then it has to be executed at sampled intervals. How does this sampling frequency affect the number of tasks that can be operational? How does the execution time for each task affect the number of tasks and sampling frequency? What is the tradeoff? How is this similar to the data acquisition system in Problem 10.6?

10.13. A DC tach generator can be replaced by a pulse mechanism, known as **quadrature feedback**. This consists of a slotted disk with two infrared detectors that are 90° apart electrically. As the disk rotates, the two detectors provide the signals, as shown in Fig. P10.13.

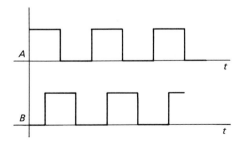

Figure P10.13 Quadrature signal output

(a) How can we tell the direction of rotation in terms of the two outputs, *A* and *B*?

(b) How can this mechanism be used as a velocity, position transducer?

(c) If our microprocessor-based control system had two high-speed counters, how could we use this to find the velocity of the rotating disk?

10.14. A common control action with digital systems is in the form of a pulse-width modulated signal. The duty cycle is 50% for zero, 100% for maximum positive, and 0% for maximum negative values of control action. Consider the pulse-width modulated signal shown in Fig. P10.14. Show that a low-pass filter, offset voltage, and gain amplifier can produce an analog output. Assume the output is to be ±10 volts.

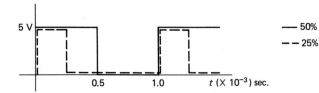

— 50%

-- 25%

Figure P10.14   Pulse-width modulated signal

10.15. Assume that we have a DC motor. If our control action is in the form of a pulse-width modulated signal, we require a signal to indicate the direction. An amplifier is provided that accepts the direction signal and uses it to redirect the pulse-width modulated signal through the motor. This provides a reversing capability. Can we put this modulated signal directly through the motor? Why? If yes, what problems or cautions are evident? The idea of filtering and frequency response will help with this problem.

10.16. Investigate the methods of D/A conversion. What are the typical times for conversion? Will this be of concern in our digital control system?

Discrete

11

The z-Transform

11.1 INTRODUCTION

In this chapter we will investigate the z-transform and its importance in the discrete domain. The important transforms will be derived, and the theorems associated with the z-transform will be investigated. The inverse z-transform will be used to return to the sampled time domain and solve difference equations. The systematic conversion of a difference equation to a computer program will be covered. This will form the basis for computer simulation. Let's begin with the definition of the z-transform.

11.2 DERIVATION OF REQUIRED Z-TRANSFORMS

This transformation should be a discrete version of the Laplace transform as given by equation (2.2.1). The integration will become a summation; our time function $f(t)$ will become a sampled version $f(kT)$, where T represents the sample time and the variable k refers to the sample in question. The exponentially decaying term in s will become a geometric series in z^{-k}. As the Laplace transform, the z-transform will be one sided and the series will converge to produce a transform in the variable z. Mathematically this is equivalent to

$$F(z) = Z[f(kT)] \stackrel{\triangle}{=} \sum_{k=0}^{\infty} f(kT) \cdot z^{-k} \qquad (11.2.1)$$

EXAMPLE 11.1   Transform of the Dirac delta ($\delta(t)$) function.

We will assume that this function has unit height at time $t = 0$. Since this function is defined at $t = 0$, we have one sample, and the z-transform becomes

$$Z[\delta(kT)] = 1 \qquad (11.2.2)$$

EXAMPLE 11.2   Transform of the step ($u(t)$) function.

The discrete version of this function is given by

$$\begin{cases} u(kT) = 0, & k < 0 \\ u(kT) = 1, & k \geq 0 \end{cases} \qquad (11.2.3)$$

and the graphical implications of the sampled function are shown in Fig.11.1.

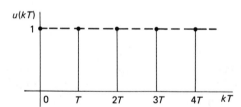

Figure 11.1   A discrete step function

SOLUTION   Using equation (11.2.1), we have

$$Z[u(kT)] = \sum_{k=0}^{\infty} 1 \cdot z^{-k} \qquad (11.2.4)$$

which generates the following geometric series:

$$Z[u(kT)] = 1 + z^{-1} + z^{-2} + z^{-3} + \cdots \qquad (11.2.5)$$

The sum of this form of geometric series is given by

$$S = \frac{A}{1 - r} \qquad (11.2.6)$$

where A is the first term in the series and r is the common factor. Using equation (11.2.6), the solution to the series and the z-transform for the step function is

$$Z[u(kT)] = \frac{1}{1 - z^{-1}} \quad \text{or} \quad \frac{z}{z-1} \qquad (11.2.7)$$

EXAMPLE 11.3   Transform of the exponentially decaying function

The discrete version of this function is given by

$$f(kT) = e^{-akT} \qquad (11.2.8)$$

and Fig. 11.2 provides the graphical implications.

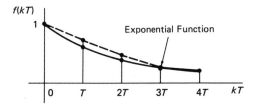

Figure 11.2   Discrete exponential decay

SOLUTION   Using equation (11.2.1), we have

$$Z[f(kT)] = \sum_{k=0}^{\infty} e^{-akT} \cdot z^{-k} \qquad (11.2.9)$$

and the corresponding geometric series:

$$Z[f(kT)] = 1 + e^{-aT} \cdot z^{-1} + e^{-a2T}z^{-2} + \cdots \qquad (11.2.10)$$

Using equation (11.2.6), we obtain the required z-transform:

$$Z[e^{-akT}] = \frac{1}{1 + e^{-aT} \cdot z^{-1}} \qquad (11.2.11)$$

EXAMPLE 11.4   Transform of sin (ωt) and cos (ωt)

If we refer to equation (2.2.13), the discrete version is given by

$$f(kT) = e^{j\omega kT} = \cos(\omega kT) + j\sin(\omega kT) \qquad (11.2.12)$$

and the z-transforms will be

$$Z[\sin(\omega kT)] = \text{Im } \{Z[e^{j\omega kT}]\} \qquad (11.2.13)$$

$$Z[\cos(\omega kT)] = \text{Re } \{Z[e^{j\omega kT}]\} \qquad (11.2.14)$$

Using equation (11.2.1), we have

$$Z[f(kT)] = \sum_{k=0}^{\infty} e^{j\omega kT} \cdot z^{-k} \qquad (11.2.15)$$

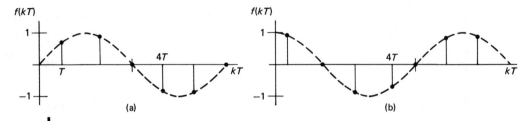

Figure 11.3   Sampled versiion of the (a) sine, (b) cosine function

and the corresponding geometric series

$$Z[f(kT)] = 1 + e^{j\omega T} \cdot z^{-1} + e^{j\omega 2T}z^{-2} + e^{j\omega 3T}z^{-3} + \cdots \qquad (11.2.16)$$

Using equation (11.2.6), we obtain the sum for the series

$$Z[f(kT)] = \frac{1}{1 - e^{j\omega T} \cdot z^{-1}} \qquad (11.2.17)$$

and, rewriting equation (11.2.7), we have

$$Z[f(kT)] = \frac{1}{1 - \cos(\omega T) \cdot z^{-1} - j\sin(\omega T) \cdot z^{-1}} \qquad (11.2.18)$$

which simplifies to

$$= \frac{1 - \cos(\omega T) \cdot z^{-1} + j\sin(\omega T) \cdot z^{-1}}{1 - 2\cos(\omega T) \cdot z^{-1} + z^{-2}} \qquad (11.2.19)$$

upon multiplication by the complex conjugate. From equations (11.2.13) and (11.2.14) we obtain the z-transform for the sine and cosine functions, respectively. That is,

$$Z[\sin(kT)] = \frac{\sin(\omega T) \cdot z^{-1}}{1 - 2\cos(\omega T) \cdot z^{-1} + z^{-2}} \qquad (11.2.20)$$

$$Z[\cos(kT)] = \frac{1 - \cos(\omega T) \cdot z^{-1}}{1 - 2\cos(\omega T) \cdot z^{-1} + z^{-2}} \qquad (11.2.21)$$

Figure 11.3 provides the sampled versions of the sine and cosine functions.

EXAMPLE 11.5   Transform of the exponentially damped sine and cosine functions.

If we refer to equation (2.2.21), the discrete version of the exponential function is

$$f(kT) = e^{-akT} \cdot e^{j\omega kT} = e^{-akT}(\cos(\omega kT) + j\sin(\omega kT)) \qquad (11.2.22)$$

and, applying similar techniques as outlined in Example 11.4, the z-transforms will emerge.   They are

$$Z[e^{-akT}\sin(\omega kT)] = \frac{(e^{-aT}\sin(\omega T))z^{-1}}{1 - 2(e^{-aT}\cdot\cos(\omega T))z^{-1} + e^{-2aT}\cdot z^{-2}} \qquad (11.2.23)$$

$$z[e^{-akT}\cos(\omega kT)] = \frac{1 - (e^{-aT}\cos(\omega T))z^{-1}}{1 - 2(e^{-aT}\cdot\cos(\omega T))z^{-1} + e^{-2aT}z^{-2}} \qquad (11.2.24)$$

As in Chapter 2, we will tabulate the derived z-transforms for future reference.   Table 11.1 provides the z-transforms.

TABLE 11.1   Properties of z-Transforms

Discrete function		z-Transform
Dirac delta	$\delta(kT)$	1
Step	$u(kT)$	$\dfrac{1}{1 - z^{-1}}$
Damped exponential	e^{-akT}	$\dfrac{1}{1 - e^{-aT}\cdot z^{-1}}$
Sine	$\sin(\omega kT)$	$\dfrac{\sin(\omega T)\cdot z^{-1}}{1 - 2(\cos(\omega T))z^{-1} + z^{-2}}$
Cosine	$\cos(\omega kT)$	$\dfrac{1 - (\cos(\omega T))\cdot z^{-1}}{1 - 2(\cos(\omega T))z^{-1} + z^{-2}}$
Damped sine	$e^{-akT}\sin(\omega kT)$	$\dfrac{(e^{-aT}\sin(\omega T))z^{-1}}{1 - 2(e^{-aT}\cos(\omega T))z^{-1} + e^{-2aT}z^{-2}}$
Damped cosine	$e^{-akT}\cos(\omega kT)$	$\dfrac{1 - (e^{-aT}\cos(\omega T))z^{-1}}{1 - 2(e^{-aT}\cos(\omega T))z^{-1} + e^{-2aT}z^{-2}}$

11.3 PROPERTIES OF Z-TRANSFORMS

Linearity

The z-transform is a linear transformation—that is,

$$Z[a_1 f_1(kT) + a_2 f_2(kT)] = a_1 Z[f_1(kT)] + a_2 Z[f_2(kT)] \qquad (11.3.1)$$

EXAMPLE 11.6

The following time function is to be sampled at 1-second intervals ($T = 1$). Find the z-transform of this function.

$$f(t) = 2 + 3e^{-t} - 3e^{-2t} \cos 6t$$

The discrete function is given by

$$f(kT) = 2 + 3e^{-kT} - 3e^{-2kT} \cos (6kT)$$

Using the linearity property, we have

$$Z[f(kT)] = Z[2] + 3 \cdot Z[e^{-kT}] - 3Z[e^{-2kT} \cos(6kT)]$$

and using Table 11.1, with $T = 1$, we obtain the z-transform:

$$Z[f(kT)] = \frac{2}{1 - z^{-1}} + \frac{3}{1 - 0.368z^{-1}} - \frac{3(1 - 0.13z^{-1})}{1 - 0.26z^{-1} + 0.018z^{-2}}$$

Advance Theorem

The **advance theorem** is similar to the derivative property of the Laplace transform. As in the derivative property, we will assume our initial conditions to be zero. The advance property allows the calculation of the z-transform of a discrete function which has been shifted n samples to the left. If the z-transform of the original discrete function is $F(z)$, then the z-transform of the function shifted n samples to the left is

$$Z[f_{k+n}] = z^n \cdot F(z) \qquad\qquad (11.3.2)$$

where $f_k = f(kT)$ and will be used throughout the text. This property will not be used and therefore is included for completion only. The reader should appreciate that a derivative in continuous time implies an advance of one sample in the discrete domain. The derivative implies future values, as does the shift to the left.

Delay Theorem

The **delay theorem** is analogous to the integral property in the Laplace domain. The delay property allows the calculation of the z-transform of a discrete function which has been shifted n samples to the right. This delay implies the past values as did the integral in the continuous sense. If the z-transform of the original discrete function is $F(z)$, then the z-transform of the function shifted

n samples to the right is

$$Z[f_{k-n}] = z^{-n} \cdot F(z) \qquad (11.3.3)$$

This property is very important and will be used extensively in the derivation of computer algorithms.

EXAMPLE 11.7

Consider the discrete function in Fig. 11.4. Find the z-transform for this function.

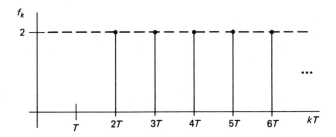

Figure 11.4   Discrete function for Example 11.7

SOLUTION   From Fig. 11.4 we find our function to be a discrete step function which has been delayed by 2 samples. The z-transform for the discrete step function is

$$\frac{2}{1 - z^{-1}}$$

and the z-transform of the delayed function is

$$Z[f_k] = \frac{2z^{-2}}{1 - z^{-1}}$$

by application of the delay theorem.

Final-Value Theorem

The **final-value theorem** will allow the calculation of the steady-state value of our output in the discrete control system. This is analogous to the final-value theorem used in the Laplace domain. The final-value theorem in the discrete domain refers to the value of the sample at infinity and is given by

$$\lim_{k \to \infty} f_k = \lim_{z \to 1} (1 - z^{-1}) \cdot F(z) \qquad (11.3.4)$$

EXAMPLE 11.8

Consider the following discrete function $Y(z)$. What is the steady-state value for the discrete time function?

$$Y(z) = \frac{z^2 + 2}{(1 - z^{-1})(z^2 - 1.6z + 0.7)}$$

SOLUTION   Using the final-value theorem, we have

$$\lim_{k \to \infty} y_k = \lim_{z \to 1} (1 - z^{-1}) \frac{z^2 + 2}{(1 - z^{-1})(z^2 - 1.6z + 0.7)}$$

and, upon simplification,

$$\lim_{k \to \infty} y_k = 30$$

These properties are the most important in the study of digital control systems. As in the Laplace domain, an inverse function must exist. Let's examine the inverse z-transform.

11.4 THE INVERSE Z-TRANSFORM

The **inverse z-transform** is attained by partial-fraction expansion. There are two cases to consider, the first containing real roots and the second containing imaginary roots in the denominator. The process is much the same as covered in the Laplace domain, differing only in form of expansion. Let's consider both cases, starting with the real roots.

Case 1

Consider the z-transform $F(z)$ as given by

$$F(z) = \frac{P(z)}{(z - p_0)(z - p_1) \cdots (z - p_n)} \qquad (11.4.1)$$

where $P(z)$ represents a polynomial in z. Referring to Table 11.1, we find the exponential decaying function having the following form:

$$\frac{z}{z - e^{-aT}}$$

after multiplication by (z/z). This implies that equation (11.4.1) could be expanded as

$$F(z) = \frac{A_0 z}{z - p_0} + \frac{A_1 z}{z - p_1} + \cdots + \frac{A_n z}{z - p_n} \qquad (11.4.2)$$

which differs from the expansion of the exponential decaying functions in the Laplace domain. The remainder of the process is the same as in the Laplace domain, and we will investigate with an example.

EXAMPLE 11.9

Consider the following z-transform. Find the discrete time function by using the inverse z-transform.

$$F(z) = \frac{z}{z^2 - 0.7z + 0.1} \qquad (1)$$

SOLUTION   Equation (1) can be written as

$$F(z) = \frac{z}{(z - 0.5)(z - 0.2)} \qquad (2)$$

Since equation (2) has real roots, we can expand as follows:

$$F(z) = \frac{A_0 z}{z - 0.5} + \frac{A_1 z}{z - 0.2} \qquad (3)$$

If we take a common denominator in equation (3) and compare with equation (2), we have

$$A_0 + A_1 = 0 \qquad (4)$$

$$-0.2\, A_0 - 0.5 A_1 = 1 \qquad (5)$$

Upon solution of the simultaneous equations and substitution of the coefficients into equation (3), we obtain

$$F(z) = \frac{3.33z}{z - 0.5} - \frac{3.33z}{z - 0.2} \qquad (6)$$

From equation (6) and Table 11.1, we obtain the required discrete time function:

$$f_k = 3.33(0.5)^k - 3.33(0.2)^k$$

The reader should study the discrete function and make sure that the process is clear.

Now let's investigate the case with imaginary roots.

Case 2

From our work in Laplace transforms it should be evident that transforms with imaginary roots in the denominator imply damped sine or cosine functions. This is also the case in the z-domain. If we generalize the form of the damped sine and cosine, we have from Table 11.1

$$\frac{A_0z^2 + A_1z}{z^2 - 2Bz + C} \tag{11.4.3}$$

where

$$e^{-aT} = \sqrt{C} \tag{11.4.4}$$

$$\omega T = \cos^{-1}\left(\frac{B}{\sqrt{C}}\right) \tag{11.4.5}$$

The numerator of equation (11.4.3) will have to be split up such that

$$\sqrt{C}\sin(\omega T)\cdot z \tag{11.4.6}$$

will be the form for a damped sine and

$$z^2 - Bz \tag{11.4.7}$$

will be the form for a damped cosine. An example will clarify this process.

EXAMPLE 11.10

Consider the following z-transform. Find the discrete time function by using the inverse z-transform.

$$F(z) = \frac{z^2 + z}{(z^2 - 1.2z + 0.7)(z - 0.6)} \tag{1}$$

SOLUTION From equation (1) we find that there are imaginary roots and one real root in the denominator. This implies an application of both case 1 and 2. Applying both cases, we have

$$F(z) = \frac{A_0z^2 + A_1z}{z^2 - 1.2z + 0.7} + \frac{A_2z}{z - 0.6} \tag{2}$$

If we solve for the coefficients, equation (2) becomes

$$F(z) = \frac{-4.706z^2 + 3.824z}{z^2 - 1.2z + 0.7} + \frac{4.706z}{z - 0.6} \tag{3}$$

Let's show equation (3) as

$$F(z) = F_1(z) + F_2(z) \tag{4}$$

Now

$$F_1(z) = \frac{-4.706z^2 + 3.824z}{z^2 - 1.2z + 0.7} \qquad (5)$$

Comparing equation (5) with (11.4.3), we have $B = 0.6$ and $C = 0.7$. Substituting into equation (11.4.5), we obtain $\omega T = 0.771$. By using equation (11.4.6) we find the numerator term for a damped sine, which is

$$0.583z \qquad (6)$$

and, using equation (11.4.7), we find the numerator term for a damped cosine, which is

$$z^2 - 0.6z \qquad (7)$$

Now equation (5) can be written as

$$F_1(z) = -4.706 \left[\frac{z^2 - 0.6z}{z^2 - 1.2z + 0.7} - \frac{0.213}{0.583} \frac{0.583z}{z^2 - 1.2z + 0.7} \right]$$

and the discrete time function can be obtained from Table 11.1:

$$f_{1k} = -4.706(0.837)^k \cos(0.771k) + 1.719(0.837)^k \sin(0.771k) \qquad (8)$$

Returning to equation (4), we have

$$F_2(z) = \frac{4.706z}{z - 0.6}$$

and from Table 11.1 we obtain the discrete time function:

$$f_{2k} = 4.706(0.6)^k$$

The total discrete time function is given by

$$f_k = (0.837)^k [1.719 \sin(0.771k) - 4.706 \cos(0.771k)] + 4.706(0.6)^k$$

It is evident that partial-fraction expansion in the z-domain requires more scrutiny. The reduction of a third-order denominator can be accommodated by the BASIC program given in Fig. 2.4. This systematic approach will save the reader a lot of hardship and will help in the process of learning.

A quick look at Chapter 2 will indicate that the ideas of discrete time convolution and dead time have been omitted in this chapter. The inverse z-transform can be attained by the discrete time convolution yet offers no simplicity over the partial-fraction method. For this reason it will not be covered. The idea of dead time in a discrete system, which is important, will be covered at a later chapter, as we require more information about the discrete domain.

11.5 DIFFERENCE EQUATIONS

As stated in Chapter 2, most physical systems can be explained in terms of differential equations with constant coefficients. In Chapter 10 we indicated that differential equations can be approximated by difference equations. This implies that the difference equation is the governing equation of the discrete control system. The *z*-transform will yield the discrete-time solution, as we will see.

EXAMPLE 11.11

Referring back to Example 2.14, we can rewrite equation (1) as

$$\frac{dv}{dt} = R\frac{di}{dt} + \frac{i}{c} \tag{1}$$

Using equation (10.2.2), equation (1) becomes

$$\frac{v_k - v_{k-1}}{T} = \frac{R(i_k - i_{k-1})}{T} + \frac{i_k}{C} \tag{2}$$

and, solving for the present current, i_k, we obtain the system difference equation:

$$i_k = \left(\frac{1}{R + \dfrac{T}{C}}\right)[v_k - v_{k-1} + R \cdot i_{k-1}] \tag{3}$$

From equation (3) we find the present current a function of the current one sample ago. If $R = 1$ MΩ, $C = 1$ μF, and $T = 0.1$ second, we can solve the difference equation. The input voltage is assumed to be 5 volts and the current values will be in microamps. Table 11.2 provides the step-by-step solution to the difference equation and compares the discrete current to the continuous current $i(t)$.

From Table 11.2 it is evident that our solution to the difference equation is approximating the continuous system. How can we find the solution to this difference equation using *z*-transforms? If we take *z*-transforms of equation (3), using the properties of *z*-transforms, we have

$$I(z) = \left(\frac{1}{R + \dfrac{T}{C}}\right)[V(z) - z^{-1}V(z) + Rz^{-1}I(z)] \tag{4}$$

TABLE 11.2 Solution to Difference Equation

k	Time (sec)	v_{k-1}	v_k	i_{k-1}	i_k	$i(t)$
0	0.0	0	5	0	4.55	5.00
1	0.1	5	5	4.55	4.13	4.52
2	0.2	5	5	4.13	3.76	4.09
3	0.3	5	5	3.76	3.41	3.70
4	0.4	5	5	3.41	3.10	3.35
5	0.5	5	5	3.10	2.82	3.03
6	0.6	5	5	2.82	2.57	2.74
7	0.7	5	5	2.57	2.33	2.48
8	0.8	5	5	2.33	2.12	2.25
9	0.9	5	5	2.12	1.93	2.03
10	1.0	5	5	1.93	1.75	1.84

and, solving for $I(z)$, we obtain

$$I(z) = \frac{V(z)(1 - z^{-1})}{R + \dfrac{T}{C}} \cdot \frac{1}{1 - \dfrac{R}{R + \dfrac{T}{C}} z^{-1}} \tag{5}$$

Our input is a step input, which yields

$$V(z) = \frac{v}{1 - z^{-1}}$$

Therefore equation (5) becomes

$$I(z) = \frac{v}{R + \dfrac{T}{C}} \cdot \frac{1}{1 - \dfrac{R}{R + \dfrac{T}{C}} z^{-1}} \tag{6}$$

and the discrete solution is given by the inverse z-transform:

$$i_k = \frac{v}{R + \dfrac{T}{C}} \cdot \left(\frac{R}{R + \dfrac{T}{C}}\right)^k \tag{7}$$

It is left as an exercise for the reader to substitute the values into equation (7) and verify the results with Table 11.2.

Let's consider another example.

EXAMPLE 11.12

An excellent example for the utilization of difference equations is the borrowing of money from financial institutions. Assume you borrow P_0 dollars at $r\%$ per term. You agree to pay u dollars per term until the loan is exhausted. What is the principal remaining after the kth term?

SOLUTION   If we require the principal after the kth payment, we are in fact requiring the principal at the $(k + 1)$ term. This is given by the following difference equation:

$$P_{k+1} = (1 + r)P_k - u \tag{1}$$

If we take z-transforms, we have

$$zP(z) - zP_0 = (1 + r)P(z) - \frac{u}{1 - z^{-1}} \tag{2}$$

Notice that equation (2) contains initial conditions and is required for this example only. If we solve for $P(z)$, we obtain

$$P(z) = \frac{P_0 z^2 - (P_0 + u)z}{(z - 1)[z - (1 + r)]} \tag{3}$$

and, using partial-fraction expansion, we have

$$P(z) = \frac{u}{r(1 - z^{-1})} + \frac{P_0 - \dfrac{u}{r}}{1 - (1 + r)z^{-1}} \tag{4}$$

The solution is given by the inverse z-transform, which is

$$P_k = \frac{u}{r} + \left(P_0 - \frac{u}{r}\right)(1 + r)^k \tag{5}$$

The reader is advised to initiate a few problems to validate equation (5).

Let's turn now to the computer solution of the difference equation, as this will provide a discrete simulation of our continuous control system.

11.6 COMPUTER SIMULATION

In Chapter 2 we simulated a differential equation with the aid of operational amplifiers. This hardware approach provided an accurate representation of the continuous system. Our differential equation is now a difference equation and requires iterative calculations for a solution. We require a software approach for our simulation. To remain language independent, we will begin with a register structure for the implementation and verify its usefulness using the BASIC language.

If we can simulate a compensation network then we can simulate a control system. This is possible, as both can be represented by a transfer function. Now equation (10.2.5) represents the difference equation for the general compensation network. From equation (10.2.5) it is evident that the storage of samples is required. Consider Fig. 11.5. This represents two separate stacks of registers, the first containing samples of the error and the second containing samples of the control action. To implement equation (10.2.5) we require constant multipliers for each register in the individual stacks. These multipliers correspond to the coefficients in equation (10.2.5). Figure 11.6 displays these multipliers as stacks of additional registers.

To calculate the present control action u_k, after every sample of error e_k, the following must occur in that order.

1. The contents of each register, in both the error and control-action stacks, is shifted upward by one position. The shifting begins at the top of each stack. This discards the information at the top and allows new information to be loaded at the bottom of each stack.

2. The present sample of the error is loaded into the bottom of the error stack.

3. Equation (10.2.5) is now implemented and the value of the present control action is generated. This control action is also loaded into the bottom of the control-action stack.

4. Return to step 1 and repeat.

The reader should understand that the continuous repetition of this process provides a solution to the discrete equation. The frequency of repetition is the sampling frequency, and the time is equivalent to the sampling time. It should be evident that a real-time operating system is required to properly implement the solution in a control system. To perform a simulation, it is performed in machine time. This will be more apparent when we look at an

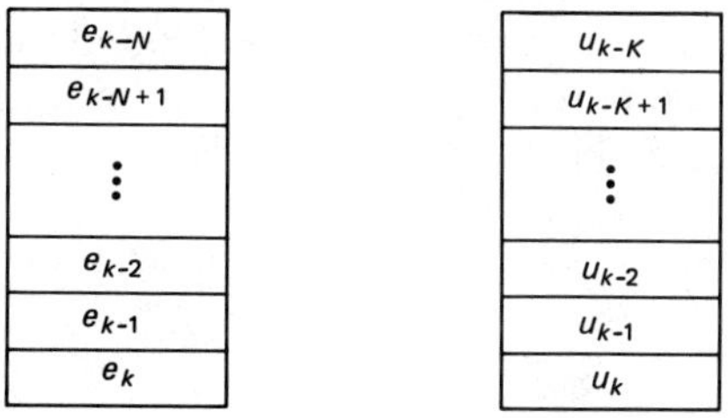

Figure 11.5 Register stack for error and control action

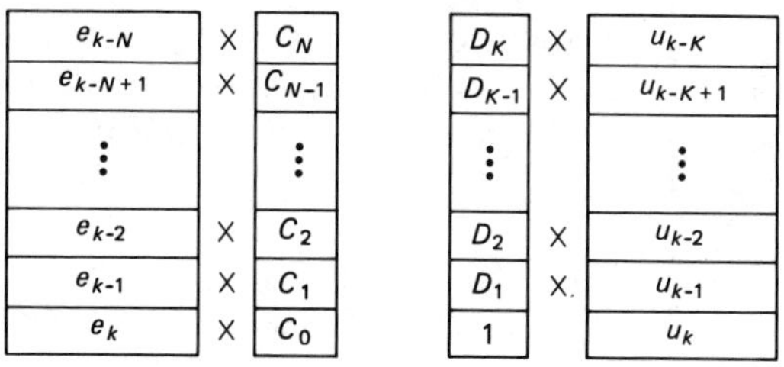

Figure 11.6 Register stacks and multipliers

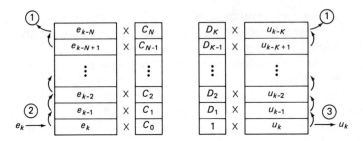

Figure 11.7 Graphical representation of steps involved in solution

example.   Figure 11.7 provides a graphical representation of the steps involved in the solution of the difference equation.   Now let's investigate the example.

EXAMPLE 11.13

Given the following difference equation, simulate the solution with a unit step input.

$$u_k = 0.0097e_k + 1.96u_{k-1} - 0.97u_{k-2} \qquad (1)$$

SOLUTION   In a high-level language the stacks in Fig. 11.7 are one dimensional arrays.   Consider the BASIC program in Fig. 11.8.

```
100   FOR K = (ORDER - 1) TO 0 STEP - 1
110       E(K + 1) = E(K)
120       U(K + 1) = U(K)
130   NEXT K
```

Figure 11.8 BASIC program to implement stacks

From Fig. 11.8 it is evident that each time the program is executed, we obtain a shifting of information in the stacks.   The variable, ORDER, represents the order of the difference equation and is equivalent to the highest number of past samples required in any variable.   In our example we require two past samples of the control action, which implies ORDER = 2.

The coefficient stacks are constant and can be loaded at the beginning of the program.   We require a unit step input, which has to be loaded into the bottom of the error stack $E(0)$.   Figure 11.9 provides the additional statements required for the program in Fig. 11.8.

```
10   C(0) = 0.0097
20   D(1) = 1.96
30   D(2) = -0.97
            . . .
150  E(0) = 1
```

Figure 11.9 Additions to BASIC program

The reader should appreciate that the first time the program is executed, the error goes from 0 to 1, and each subsequent pass remains at 1. This is representative of sampling a unit step function. Now we can generate the present control action, which is given by equation (1). Figure 11.10 provides the BASIC statement to implement the present control action.

```
160 U(0) = C(0)*E(0) + D(1)*U(1) + D(2)*U(2)
```

Figure 11.10   BASIC implementation of present control action

If we combine the statements from Figs. 11.8 through 11.10, we obtain the complete program to implement the difference equation. If this program is initiated every T seconds, then it will be running in real time. In the case of the simulation, we will insert the program into a loop. Every pass through the loop represents another sample. As the speed of processing is dependent on the machine, this is considered to be running in machine time.

Figure 11.11 provides the complete simulation program. The difference equation represents a sampling time of 0.1 second and is embedded in equation (1). We will investigate the generation of the difference equation from a transfer function in the next chapter, and for now we accept the difference equation as given.

```
10     C(0) = 0.0097
20     D(1) = 1.96
30     D(2) = -0.97
90     FOR I = 1 TO 50
100        FOR K = (ORDER - 1) TO 0 STEP -1
110            E(K + 1) = E(K)
120            U(K + 1) = U(K)
130        NEXT K
150    E(0) = 1
160    U(0) = C(0)*E(0) + D(1)*U(1) + D(2)*U(2)
170    PRINT "Control Action = ";U(0),"Time = "*0.1
180    NEXT I
190    END
```

Figure 11.11   Complete BASIC simulation program

The reader is encouraged to run the program as given in Fig. 11.11 and compare the results with Problem 10.2. What if we would like to simulate a complete control system? Consider the control system as shown in Fig. 11.12. It is evident that we require another stack of registers to contain past samples of the system output. We also require a stack of registers for the constant multipliers contained in the plant difference equation, G_M. Figure 11.13 provides the register implementation of the control system.

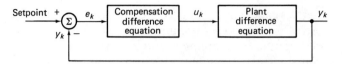

Figure 11.12   Control system for simulation

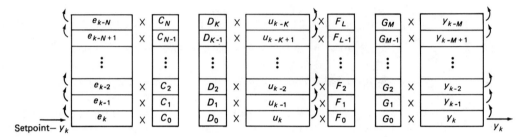

Figure 11.13   Register implementation of control system

From Fig. 11.13 we can generate the BASIC program to simulate the control system.   The BASIC program is given in Fig. 11.14.

```
10    C(*) = *
40    D(*) = *
50    F(*) = *
70    G(*) = *
90    FOR I = 1 TO 50
100       FOR K = (ORDER - 1) TO 0 STEP -1
110           E(K + 1) = E(K)
120           U(K + 1) = U(K)
130           Y(K + 1) = Y(K)
140       NEXT K
150   E(0) = SETPOINT - Y(0)
160   U(0) = COMPENSATION DIFFERENCE EQUATION
170   Y(0) = PLANT DIFFERENCE EQUATION
180   PRINT U(0),Y(0)
190   NEXT I
200   END
```

Figure 11.14   BASIC program to simulate control system

EXAMPLE 11.14

Given the following compensation and plant difference equations.   If the input is a unit step and the sampling time is 0.1 second, generate the program to simulate the unity feedback control system.

$$u_k = 2 \cdot e_k$$

$$y_k = 0.0097u_k + 1.96y_{k-1} - 0.97y_{k-2}$$

SOLUTION   The required BASIC program is easily obtained by modifying Fig. 11.14 as shown in Fig. 11.15.

```
10      C(0) = 2
40      F(0) = 0.0097
70      G(1) = 1.96
80      G(2) = -0.97
90      FOR I = 1 TO 50
100         FOR K = 1 TO 0 STEP -1
110             E(K + 1) = E(K)
120             U(K + 1) = U(K)
130             Y(K + 1) = Y(K)
140         NEXT K
150     E(0) = 1 - Y(0)
160     U(0) = C(0)*E(0)
170     Y(0) = F(0)*U(0) + G(1)*Y(1) + G(2)*Y(2)
180     PRINT U(0),Y(0),"Time =";I*0.1
190     NEXT I
200     END
```

Figure 11.15   BASIC program for Example 11.14

The reader is encouraged to run the program in Fig. 11.15 and plot the control action and system output with respect to time. The system output is assumed to be continuous and should therefore be a smooth curve. The control action represents the output of a D/A converter at sampled intervals and should therefore be discrete in nature. In the next chapter we will investigate the discrete transfer function.

11.7 SUMMARY

In this chapter we introduced the z-transform and derived the common transforms of discrete time functions. We investigated the properties of z-transforms and the method of partial-fraction expansion for the inverse z-transform. We turned our attention to difference equations and used z-transforms for the solution of the same. The reader was referred back to Chapter 2 in the hope of recognizing the similarity of the z-transform to the Laplace transform.

The solution of the difference equation was investigated via computer. A language-independent structure was developed and verified by using the BASIC language. The simulation of a compensating network led to the simulation of the entire control system.

Problems

11.1. Using the definition of the z-transform, generate the z-transforms for the discrete damped sine and cosine functions.

11.2. A ramp function is normally used in a control system to provide a gradual setpoint change. Figure P11.2 provides the graphical implications of the ramp function. Mathematically it is given by

$$f_k = kT$$

Show that the z-transform for the ramp function is

$$F(z) = \frac{Tz^{-1}}{(1 - z^{-1})^2}$$

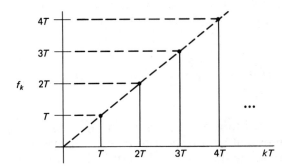

Figure P11.2 Discrete ramp function

11.3. Figure P11.3 represents the typical use of the ramp function in a setpoint change. Show that the z-transform for this function is

$$F(z) = \frac{T(z^{-1} + z^{-2})}{(1 - z^{-1})}$$

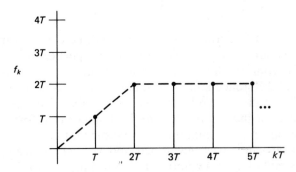

Figure P11.3 Modified discrete ramp function for Problem 11.3

11.4. Using Table 11.1 (page 199), $T = 0.1$ second, find the z-transforms for the following discrete time functions.

(a) $f_k = e^{-3kT}$ (b) $f_k = 2 \cos (3kT)$

(c) $f_k = 2e^{-6kT} \cos (6kT)$ (d) $f_k = \cos (3kT) \sin (3kT)$

(e) $f_k = 3e^{-0.5kT} [\cos (2kT) - 2 \sin (2kT)]$

11.5. Given the following discrete time function, find the z-transform if $T = 1$ second. Now find the inverse z-transform. Is the new discrete time function the same as the original? Why?

$$f_k = 6 \cos (6kT)$$

11.6. Use partial-fraction expansion to find the inverse z-transforms for the following z functions.

(a) $F(z) = \dfrac{z^2 + z}{(z - 0.5)(z^2 - 1.1z + 0.5)}$

(b) $F(z) = \dfrac{z^2 + 2z}{(z - 0.6)(z - 0.3)(z - 0.1)}$

(c) $F(z) = \dfrac{z^2 + 3z}{(z^3 - 1.6z^2 + 1.8z - 2.4)}$

11.7. Apply the final-value theorem to the z-transforms given in Problems 11.2 and 11.3. Are they in agreement with the graphical predictions?

11.8. Using equations (10.2.2) and (10.2.3), generate the difference equation for the physical system given in Problem 2.12. Assume $T = 0.1$ second. Find the solution by brute-force calculation and then via z-transforms. How does it compare with the continuous solution found in Problem 2.12? Repeat this question with $T = 0.05$ second.

11.9. You borrow $10,000 for 3 years. The interest rate is set at 12% per annum. Using Example 11.12 as a reference, what are the monthly payments? Generate a computer program to calculate the monthly interest and principal payments.

11.10. You borrow $20,000 at 15% per annum. You can afford to pay $456.78 per month. How many months will it take to pay off the loan?

11.11. Consider the following difference equation. Generate a computer simulation with a unit step input. The sampling time is 0.1 second. Plot the points. Is this function familiar?

$$u_k = 0.1e_{k-1} + 2u_{k-1} - u_{k-2}$$

11.12. Consider the following difference equation. Generate the computer simulation with a unit step input. The sampling time is 0.1 second. Plot the points. Is this function familiar?

$$u_k = 0.1e_{k-1} + 0.1e_{k-2} + u_{k-1}$$

11.13. Repeat Problems 11.11 and 11.12 if the inputs are the Dirac delta or unit impulse functions. This function is implemented by loading the error stack with 1 on the first pass and loading 0's on subsequent passes.

11.14. Consider the control system given in Fig. P11.14. This represents a lag compensation network with a second-order plant.

    (a) Use equations (10.2.2) and (10.2.3) to generate the difference equations for the compensating network and plant. Leave sampling time T as a variable. Generate the computer simulation for various values of T, $0.1 < T < 5$. Plot the output and control action for all cases and comment on the role of sampling time.

    (b) If $T = 0.1$ second, investigate the response with the location of the zero moving in the lag compensator.

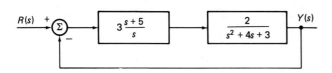

Figure P11.14   Control system for Problem 11.14

11.15. If the compensating network in Problem 11.14 is replaced with the following lead compensator,

$$D(s) = \frac{4s}{s + 6}$$

    (a) Repeat 11.14(a).

    (b) Repeat 11.14(b) with the location of the pole varying.

11.16. Refer to Problem 2.15. Provide a computer simulation for this plant with the various values of c. Use a sampling time of $T = 0.1$ second. Compare the outputs with those obtained in Problem 2.15.

Part II
Discrete

12

Discrete Transfer Functions

12.1 INTRODUCTION

The definition of the transfer function remains as outlined in Chapter 3. In this chapter we will investigate the discrete transfer function. We will generate transfer functions from difference equations and vice versa. The z-plane will be introduced and the locations of poles and zeroes investigated. These locations will be compared to the locations on the s-plane. The frequency response of a system or component will be investigated with an emphasis on the importance of sampling time.

12.2 TRANSFER FUNCTION FROM DIFFERENCE EQUATION

We will consider difference equations which contain past samples only. This implies the application of the delay theorem as outlined in Chapter 11. Consider the following three steps.

Step 1 Take z-transforms of both sides of the difference equation by applying the delay theorem.

Step 2 Solve for the output/input relationship in terms of z.

Step 3 Multiply numerator and denominator by the highest positive order of z.

EXAMPLE 12.1

Consider the following difference equation. Find the discrete transfer function $U(z)$.
$$u_k = u_{k-1} + 0.3e_k - 0.4e_{k-1}$$
SOLUTION Taking z-transforms, we have
$$U(z) = z^{-1}U(z) + 0.3E(z) - 0.4z^{-1}E(z)$$
and, solving for $U(z)/E(z)$, we obtain
$$\frac{U(z)}{E(z)} = \frac{0.3 - 0.4z^{-1}}{1 - z^{-1}}$$
Upon simplification, we obtain
$$\frac{U(z)}{E(z)} = \frac{0.3z - 0.4}{z - 1}$$

The discrete transfer function will allow analysis and design in the discrete frequency domain. To implement the transfer function via computer, we would require the discrete time difference equation from the discrete transfer function. Let's turn our attention to this inverse operation.

12.3 DIFFERENCE EQUATION FROM TRANSFER FUNCTION

This process involves three steps:

Step 1 Multiply numerator and denominator of the discrete transfer function by the highest negative power of z.

Step 2 Cross multiply and take inverse z-transforms.

Step 3 Solve for the present output in terms of past samples of the output and input.

EXAMPLE 12.2

Consider the following discrete transfer function. Provide a difference equation to allow computer implementation.

$$\frac{U(z)}{E(z)} = \frac{z^2 + z}{z^2 - 1.2z + 0.5}$$

SOLUTION   Multiplying by z^{-2}, we obtain

$$\frac{U(z)}{E(z)} = \frac{1 + z^{-1}}{1 - 1.2z^{-1} + 0.5z^{-2}}$$

Cross multiplying and taking the inverse z-transform, we have

$$u_k - 1.2u_{k-1} + 0.5u_{k-2} = e_k + e_{k-1}$$

and, solving for the present output, we obtain the required difference equation, which is

$$u_k = e_k + e_{k-1} + 1.2u_{k-1} - 0.5u_{k-2}$$

Let's investigate the z-plane and compare pole locations on this plane with those on the s-plane. This allows the application of theory from the continuous system and minimizes duplication.

12.4 z-PLANE AND POLE LOCATIONS

If we refer to Table 2.1 and Table 11.1, we can provide the pole locations in the s-plane and z-plane. Table 12.1 provides this information.

TABLE 12.1   Pole Locations on the s- and z-planes.

	Pole locations	
Function	s-Plane	z-Plane
Step	$s = 0$	$z = 1$
Damped exponential	$s = -a$	$z = e^{-aT}$
Sine, cosine	$s_{1,2} = \pm j\omega$	$z_{1,2} = \cos(\omega T) \pm j \sin(\omega T)$
Damped sine, cosine	$s_{1,2} = -a \pm j\omega$	$z_{1,2} = e^{-aT}[\cos(\omega T) \pm j \sin(\omega T)]$

From Table 12.1 it is evident that the s- and z-plane poles are related by

$$z = e^{sT} \tag{12.4.1}$$

The location of the zeroes is not necessarily correct with respect to equation 12.4.1. The location of the poles dictates system stability and it is therefore sufficient information. The reader should appreciate this mechanism of travelling between the s- and z-planes and use this information for the design and analysis of digital control systems. In the first part of this text we investigated the classical control theory as related to continuous control systems. This information can now be carried over via the relationship given by equation (12.4.1).

On the s-plane we found that a system was stable if the poles remained on the LHP. A system was marginally stable if the poles were located on the imaginary axis. If the poles were located on the RHP, then the system was considered to be unstable. How do we define these regions on the z-plane? If we consider the pure sine or cosine function, we will isolate the region of marginal stability on the z-plane. From Table 12.1 the pole locations for the sine, cosine function are

$$z_{1,2} = \cos(\omega T) \pm j \sin(\omega T)$$

which in polar form is

$$z_{1,2} = 1 \angle \pm \omega T \tag{12.4.2}$$

From equation (12.4.2) it is evident that the pole locations for all values of ωT lie on a unit circle. The boundary for stable and unstable systems on the z-plane is the unit circle. This is equivalent to the imaginary axis on the s-plane. With a quick glance at Table 12.1, it is evident that stable systems have poles inside the unit circle. This leaves the area outside the unit circle for unstable systems. Figure 12.1 portrays this information.

The reader should be wondering about the implications of sampling time on the location of the poles. What is the maximum sampling time allowed?

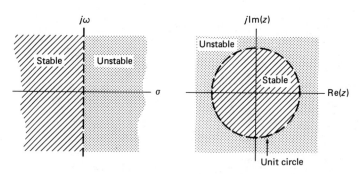

Figure 12.1   Stability on the s- and z-planes

Problem 11.5 hinted at this question, and we will see how your answer compares with reality.

12.5 SAMPLING TIME

We have found a relationship between the locations of poles on the s- and z-planes. This relationship contained the variable T, or sampling time. If we consider the pure sine or cosine function, we find, from equation (12.4.2), that the pole locations are unique if

$$\omega T \leq \pi \qquad (12.4.3)$$

now

$$T = \frac{2\pi}{\omega_s}; \qquad \omega_s \rightarrow \text{sampling frequency} \qquad (12.4.4)$$

which yields

$$\omega_s \geq 2 \cdot \omega \qquad (12.4.5)$$

From equation (12.4.5) it is evident that our sampling frequency has to be at least twice the frequency of the sampled function. This represents the sampling theorem and can be stated formally as follows:

If a signal is to be reconstructed from samples, then it should be sampled at a frequency that is at least twice the value of the highest-frequency component in the signal.

This provides a guideline for proper selection of the sampling time. Practically, the sampling frequency is four to ten times the highest frequency content. The reason for this will become apparent in the next section, as we investigate the frequency response of discrete functions.

EXAMPLE 12.3

A continuous system has the following closed-loop transfer function. This system is to be controlled via computer. What is the required sampling time (T)?

$$P(s) = \frac{s + 2}{s^2 + 0.2s + 1.01}$$

SOLUTION   If we rewrite the transfer function as

$$P(s) = \frac{s + 2}{(s + 0.1)^2 + 1}$$

we find the frequency content to be

$$\omega = 1 \text{ rad/sec}$$

This implies that our sampling frequency should be at least 2 rad/sec. Practically we would choose a sampling frequency of at least 8 rad/sec. This is equivalent to a sampling time of 0.8 second. Any sampling time less than 0.8 second would be acceptable, as this would approximate the continuous system more accurately.

In multitasking systems it is important to sample the proper amount and not to oversample. Oversampling requires computer time—time that could be spent on another task. Consider the following example.

EXAMPLE 12.4

A microprocessor-based control system is multitasking and can handle up to eight tasks in real time. A discrete lead compensator is to be used which requires 20 msec of execution time.

(a) If all eight tasks are to perform the lead compensation on eight identical plants, what is the minimum sampling time allowed for any of the tasks?

(b) If the sampling time required to control a plant is 50 msec, how many plants can we control with our multitasking system? Is any computer time available? How can it be used?

SOLUTION It is assumed that the 20-msec execution time contains the A/D conversion, solution to difference equation, and D/A conversion for each task. This implies that eight tasks require 0.16 second of computer time. The minimum sampling time is therefore 0.16 second, as the computer can guarantee a real-time solution for each task within that time. The computer is running at 100% usage, which in practical terms would not be acceptable. If each task required 21 msec for execution, the computer could no longer provide the real-time solution. The RTOS would acquire an overlap error and terminate the execution of all tasks. To avoid this problem, system designers specify a maximum computer usage. If our system were to be operating at 80% usage, the minimum sampling time would be 0.20 second. This would allow for any changes in execution times for the tasks and maintain a real-time solution.

The reader should appreciate that our control system contained identical plants. The calculation of computer usage for different plants with different sampling times is more involved. Some manufacturers provide an indication of computer usage on the computer system to simplify the process. The answer to part (a), in practical terms, would be approximately 0.2 second.

In part (b) we find that only two tasks could run in real time. This provides an 80% computer-time usage, since we have 10 msec remaining between each sample. Our system is capable of running eight tasks. The balance of the tasks could be running in the background on machine time left over from the real-time tasks. These background tasks can be used for operator interface elements such as keyboards and screens.

The preceding example provided insight to the complexity of RTOS with multitasking capabilities. The computer, which is considered to be fast, can have problems closing the loop on high-speed control systems. In such systems a dedicated microprocessor-based control system is utilized. Another problem incurred by sampling is readily visible in the frequency response of the discrete system. We turn our attention to this area.

12.6 FREQUENCY RESPONSE OF DISCRETE FUNCTIONS

In the continuous system, the frequency response of a system was attained by replacing the Laplace operator by $j\omega$. The magnitude and phase were then plotted with respect to frequency. If we consider a discrete transfer function $P(z)$ and equation (12.4.1), we have the frequency response given by

$$P(e^{j\omega T}) \tag{12.6.1}$$

The DC gain would be given by

$$P(1) \tag{12.6.2}$$

since $s = 0$.

EXAMPLE 12.5

Consider the following discrete transfer function $P(z)$. What is the DC gain for the function? Provide the expression for the magnitude of the transfer function. Sketch the magnitude with respect to frequency. Is there any problem? The transfer function reflects a sampling time of 1 second or a sampling frequency of 6.28 rad/sec.

$$P(z) = \frac{0.431(z + 0.935)}{z^2 + 0.985z + 0.819}$$

SOLUTION   The DC gain is given by $P(1)$ and is equal to 1. The frequency response is given by

$$P(e^{j\omega T})_{T=1} = \frac{0.431(e^{j\omega} + 0.935)}{e^{j2\omega} - 0.985e^{j\omega} + 0.819}$$

and the magnitude is

$$|P(e^{j\omega T})_{T=1}|$$

$$= \frac{\sqrt{[0.403 + 0.431 \cos (\omega)]^2 + [0.431 \sin (\omega)]^2}}{\sqrt{[0.819 + \cos (2\omega) - 0.985 \cos (\omega)]^2 + [\sin (2\omega) - 0.985 \sin (\omega)]^2}}$$

with the phase being

$$\angle P(e^{j\omega T})_{T=1} = \arctan \left(\frac{0.431 \sin (\omega)}{0.403 + 0.431 \cos (\omega)} \right)$$

$$- \arctan \left(\frac{\sin (2\omega) - 0.985 \sin (\omega)}{0.819 + \cos (2\omega) - 0.985 \cos (\omega)} \right)$$

The reader is asked to generate a computer program and plot the magnitude and phase with respect to frequency. From the accurate plots, the reader is asked to compare to the results of Problem 3.8. The transfer function represents a low-pass filter with a bandwidth (*BW*) of 1.5 radians. Figure 12.2 provides an ideal magnitude response of the transfer function and should approximate the accurate plots.

From Fig. 12.2 we find the discrete function providing a lowpass function from DC to the bandwidth. Unfortunately, this spectrum is repeated at integer multiples of the sampling frequency. This implies that higher frequencies will be included in our lowpass function. This presents a problem in that practical systems contain noise which are of higher frequencies than the system bandwidth. If we could guarantee a band-limited signal, then our digital lowpass filter would operate similar to the continuous type.

To limit the frequency content of our signal, we would prefilter the signal before digitizing it. This implies the addition of a continuous or analog lowpass filter. The order of the filter would be dictated by the sampling frequency. From Fig. 12.2 it is evident that if the sampling frequency approached infinity, the analog prefilter would no longer be required. In the preceding section we indicated that oversampling wasted computer time, and so a tradeoff exists. The reader should be convinced that an all-pass function exists in Fig. 12.2 if our sampling frequency is less than twice the bandwidth of our system. This agrees with the sampling theorem as stated previously.

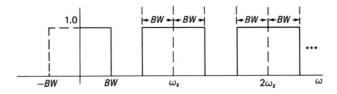

Figure 12.2   Ideal magnitude plot for discrete transfer function

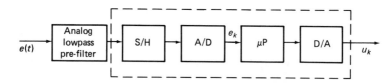

Figure 12.3   Modified digital version of compensating network

The preceding example indicated the problems with frequency response due to sampling. Since our compensation network is a transfer function, it will suffer the same limitations. The discrete version of our compensation network, as shown in Fig. 10.6, should be modified as per Fig. 12.3. This will provide a practical implementation of the compensating function.

We have all the components necessary to close the loop with a computer. In the next chapter we will investigate the analysis of closed-loop control systems with the computer providing the compensation.

12.7 SUMMARY

In this chapter we generated the discrete transfer function from a difference equation and vice versa. The z-plane was introduced and the pole locations from the s-plane were mapped onto the z-plane via e^{sT}. The unit circle was found to be the boundary for stable and unstable systems. In terms of sampling frequencies, we found that a minimum sampling frequency existed. This was equivalent to twice the highest frequency content of the signal. Formally this was known as the *sampling theorem*. In practical systems, the sampling frequency would be taken as four to ten times the highest frequency content of the signal. Example 12.4 provided an insight into the tradeoffs incurred in an RTOS with multitasking and sampling times.

We investigated the frequency response of discrete transfer functions. It was evident that the magnitude response repeated itself at integer intervals of the sampling frequency. To limit this effect, an analog lowpass prefilter was required. The order of this filter was related to the sampling frequency. If a higher sampling frequency was used, a lower order of filter would suffice. This presented another tradeoff for the reader to consider.

Problems

12.1. Provide the transfer functions for the difference equations given in Problems 11.11 and 11.12.

12.2. Provide the difference equations for the transfer functions given in Problems 11.2 and 11.3.

12.3. Refer to Table 12.1. Plot the migration of the pole for the damped exponential function as T varies from 0 to infinity. Assume $a = 1$. What is the significance of the migration on the z-plane?

12.4. Refer to Table 12.1. Plot the migration of the poles for the damped sine function as T varies from 0 to infinity. Assume $a = 1$ and $\omega = 2$ rad/sec. What is the significance of the migration on the z-plane? What is the maximum sampling time to guarantee that the pole locations are unique?

12.5. If a system has zeroes outside the unit circle, is it considered to be unstable? Why?

12.6. From the sampling theorem we find that our signal cannot contain frequency components that are greater than one-half the sampling frequency (ω_s). This area is shown on the s-plane in Fig. P12.6. The area in question is known as the **primary strip**. If the system poles stay within the primary strip, then our sampling time is sufficient. Given the following transfer functions, find the corresponding sampling times to guarantee that the poles remain in the primary strip.

(a) $P(s) = \dfrac{s}{s^2 + 2s + 16}$

(b) $P(s) = \dfrac{s + 3}{s^2 + 2s + 40}$

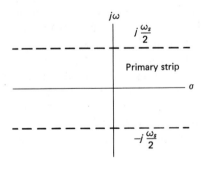

$j\omega$

$j\dfrac{\omega_s}{2}$

Primary strip

σ

$-j\dfrac{\omega_s}{2}$

Figure P12.6   Primary strip on the s-plane

12.7. What are the practical sampling times required for the systems given in Problem 12.6?

12.8. Consider the first-order transfer function given below. A practical sampling time is given by one-fifth the system time constant.   What would be the required sampling time?

$$P(s) = \frac{2}{s + 8}$$

12.9. Consider the following transfer function.   What would be the required sampling time?   Problem 3.10 will help.

$$P(s) = \frac{2}{s^2 + 4s + 3}$$

12.10. Consider the following difference equation.   Generate a program in machine language (microprocessor of your choice) to implement the difference equation.   Calculate the computation time required for your algorithm.   Assume the overhead for the A/D and D/A conversion is 50 microseconds.   If the microprocessor was servicing only this algorithm, what would be the maximum sampling frequency possible?

$$u_k = e_k + 0.5e_{k-1} - u_{k-1}$$

12.11. How would the sampling frequency change in Problem 12.10 if the difference equation were given as follows?

$$u_k = e_k + 0.31e_{k-1} - 0.61u_{k-1}$$

12.12. A real-time multitasking computer control system can be generated by using time-driven interrupts.   Is this true?   If so, incorporate programs generated in Problems 12.10 and 12.11 such that the first is executed every 0.1 second and the second every 0.2 second.

12.13. Provide a computer simulation of $P(z)$ as given in Example 12.5.   The input is a unit step function.   Is it a lowpass filter?

12.14. The transfer function $P(z)$ for Example 12.5 can be shown as

$$P(z) = \frac{Dz + E}{z^2 - 2Az + B}$$

where

$$A = e^{-0.1T} \cos (0.995T)$$

$$B = e^{-0.2T}$$

$$C = e^{-0.1T} \sin (0.995T)$$

$$D = 1 - A - 0.1005C$$

$$E = -A + B + 0.1005C$$

Generate an expression for the magnitude of the transfer function $P(z)$. Leave the sampling time (T) as a variable.   Generate a computer program to plot the magnitude with respect to frequency with the sampling time varying from 0.4 to 4 seconds. What is the result?

12.15. Aliasing is a term given to the point where higher frequencies are folded into lower frequencies. From Problem 12.14, what is the sampling frequency at which aliasing begins? How does this frequency compare with the bandwidth of the system?

12.16. Figure P12.16 provides the magnitude of a continuous lowpass second-order filter. If this filter is used as a prefilter for the system in Example 12.5, what is the required sampling frequency for the digital system such that high-frequency content is eliminated?

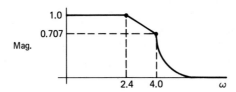

Figure P12.16   Magnitude of lowpass filter for Problem 12.16

12.17. Consider the following continuous lag compensator. If this compensator is to be digitally generated, what would be the required sampling frequency? Consider the bandwidth using Bode analysis.

$$D(s) = \frac{s + 1}{s + 0.1}$$

Discrete

13

Discrete Control System Analysis

13.1 INTRODUCTION

If we are to analyze the performance of our digital compensation network, then we have to be able to obtain the response of our continuous plant relative to a discrete control action. The most common process involves the *zero-order hold process* (*ZOH*). We will develop this process and verify its accuracy with examples. This process will generate $G(z)$ and will allow the analysis of the closed-loop control system in the discrete sense. The closed-loop transfer function and characteristic equation can be generated. The pole locations can be plotted on the z-plane and the discrete root locus can be generated. We will concentrate on a proportionally controlled second-order plant. This will provide manageable mathematics and a logical basis for comparison to our study of the second-order plant in the continuous sense.

13.2 *G(z)* FROM A CONTINUOUS PLANT DRIVEN BY ZERO-ORDER HOLD

From Chapter 10 we found that the output of our discrete compensation net-work was obtained from a D/A converter. The control action (u_k) would be passed to the D/A converter, which in turn would provide a continuous staircase signal $u(t)$. This process is known as **zero-order hold**. Figure 13.1 portrays the process.

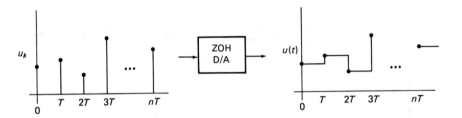

Figure 13.1   ZOH process

From Figure 13.1 we can show the control action as per Fig. 13.2. This is equivalent to an infinite sum of rectangular pulses. If we find the output to each rectangular pulse input, then the system output will be obtained from the algebraic sum of the individual outputs. If we consider the nth rectangular pulse in Fig. 13.2, this can be represented by two step functions as indicated in Fig. 13.3. If our continuous plant has a transfer function equal to $G(s)$, then the output is

$$u_n \cdot \mathscr{L}^{-1}\left[\frac{G(s)}{s}\right] \tag{13.2.1}$$

relative to a step function u_n/s. The discrete output is

$$u_n Z\left\{\mathscr{L}^{-1}\left[\frac{G(s)}{s}\right]\right\} \tag{13.2.2}$$

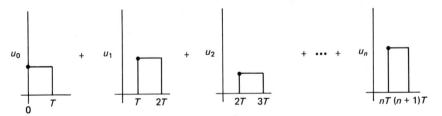

Figure 13.2   Separation of ZOH output

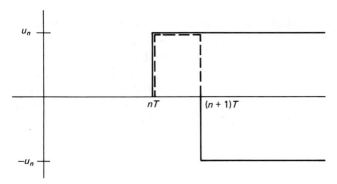

Figure 13.3   The *n*th rectangular pulse for summing two step functions.

From Fig. 13.3 we find our positive and negative step functions delayed by n and $n + 1$ samples, respectively. Therefore our discrete output due to the nth rectangular pulse is

$$Y_n(z) = u_n z^{-n} Z \left\{ \mathcal{L}^{-1} \left[\frac{G(s)}{s} \right] \right\} - u_n z^{-(n+1)} Z \left\{ \mathcal{L}^{-1} \left[\frac{G(s)}{s} \right] \right\} \qquad (13.2.3)$$

Simplifying, we have

$$Y_n(z) = u_n z^{-n} (1 \cdot z^{-1}) \cdot Z \left\{ \mathcal{L}^{-1} \left[\frac{G(s)}{s} \right] \right\} \qquad (13.2.4)$$

Now our total system output is given by

$$Y(z) = \sum_{n=0}^{\infty} Y_n(z) \qquad (13.2.5)$$

which yields

$$Y(z) = \left(\sum_{n=0}^{\infty} u_n z^{-n} \right) (1 - z^{-1}) Z \left\{ \mathcal{L}^{-1} \left[\frac{G(s)}{s} \right] \right\} \qquad (13.2.6)$$

On close inspection of equation (13.2.6), we find the summation providing the z-transform of a function u_n. Therefore equation (13.2.7) becomes

$$Y(z) = U(z) \cdot (1 - z^{-1}) \cdot Z \left\{ \mathcal{L}^{-1} \left[\frac{G(s)}{s} \right] \right\} \qquad (13.2.7)$$

or

$$\frac{Y(z)}{U(z)} = G(z) = (1 - z^{-1}) Z \left\{ \mathcal{L}^{-1} \left[\frac{G(s)}{s} \right] \right\} \qquad (13.2.8)$$

which provides an output-to-input relationship, or $G(z)$. Equation (13.2.8) is very important in that it allows the translation of a continuous transfer function into a discrete transfer function. From the discrete transfer function a computer implementation can be obtained from the governing difference equation.

Our derivation was based on a continuous plant, and it should be clear that it applies to any continuous transfer function. This implies that our compensating networks in the continuous sense can now be transferred into the discrete domain. The reader should be aware that ZOH is one of many methods that we will investigate in this text. In Chapter 10 we approximated the continuous transfer function by approximating the governing differential equation, and it would be useful to compare the results.

EXAMPLE 13.1

If we refer to Example 10.1, we find the compensation network to be

$$D(s) = \frac{1}{s + 2}$$

From Problem 12.8 we choose the sampling time to be 0.1. Equation (13.2.8) can be written as

$$D(z) = (1 - z^{-1})Z\left\{\mathcal{L}^{-1}\left[\frac{D(s)}{s}\right]\right\} \qquad (1)$$

to reflect the conversion of $D(s)$ to $D(z)$. With reference to equation (1), we have

$$\mathcal{L}^{-1}\left[\frac{D(s)}{s}\right] = \mathcal{L}^{-1}\left[\frac{1}{s(s + 2)}\right]$$

which is

$$\mathcal{L}^{-1}\left[\frac{D(s)}{s}\right] = 0.5 - 0.5e^{-2t}$$

Now

$$Z\left\{\mathcal{L}^{-1}\left[\frac{D(s)}{s}\right]\right\} = \frac{0.5}{1 - z^{-1}} - \frac{0.5}{1 - 0.819z^{-1}}$$

for $T = 0.1$ second. Multiplying by $(1 - z^{-1})$ and simplifying, we obtain

$$D(z) = \frac{0.0906}{z - 0.819}$$

The governing difference equation is given by

$$u_k = 0.0906e_{k-1} + 0.819u_{k-1}$$

If we simplify equation (10.2.8), $T = 0.1$, we have

$$u_k = 0.083e_k + 0.833u_{k-1}$$

We find the two difference equations very similar.   The reader is asked to provide a computer simulation for the difference equation obtained from ZOH and compare the results with Fig. 10.5.

Let's investigate the application of ZOH on a second-order plant.

EXAMPLE 13.2

Consider the following second-order plant.   Decide on a proper sampling time and generate $G(z)$ using ZOH.

$$G(s) = \frac{2}{s^2 + 4s + 3}$$

SOLUTION   If we use the expression for bandwidth as given in Problem 3.10, we find this system has a bandwidth of 0.9 radian.   If we take the sampling frequency at seven times the bandwidth, a sampling time of 1 second results.   How does this compare to your results for Problem 12.9?   With reference to equation (13.2.8), we have

$$\mathcal{L}^{-1}\left[\frac{G(s)}{s}\right] = \frac{2}{s(s + 3)(s + 1)}$$

which is

$$\mathcal{L}^{-1}\left[\frac{G(s)}{s}\right] = 0.667 + 0.333e^{-3t} - e^{-t}$$

Now

$$Z\left\{\mathcal{L}^{-1}\left[\frac{G(s)}{s}\right]\right\} = \frac{0.667}{1 - z^{-1}} + \frac{0.333}{1 - 0.0498z^{-1}} - \frac{1}{1 - 0.368z^{-1}}$$

for $T = 1$ second.   Multiplying by $(1 - z^{-1})$ and simplifying, we obtain

$$G(z) = \frac{0.316z + 0.0849}{z^2 - 0.418z + 0.0183}$$

The preceding example contained real pole locations, and it would be interesting to consider a second-order plant with imaginary pole locations.

EXAMPLE 13.3

Consider the following second-order plant.   Decide on a proper sampling time and generate $G(z)$ using ZOH.

$$G(s) = \frac{1}{s^2 + 0.2s + 1}$$

SOLUTION   The plant has poles with a frequency component of approximately 1 rad/sec.   If we sample at 1 second, then the poles will remain in the primary strip.   If we consider the bandwidth expression from Example 3.10, we find this plant has a bandwidth of 1.5 radians.   The 1-second sampling time still guarantees a sampling frequency which is four times the bandwidth.   Therefore the 1-second sampling time is sufficient.   With reference to equation (13.2.8), we have

$$\mathcal{L}^{-1}\left[\frac{G(s)}{s}\right] = \mathcal{L}^{-1}\left[\frac{1}{s[(s + 0.1)^2 + 0.99]}\right]$$

which is

$$\mathcal{L}^{-1}\left[\frac{G(s)}{s}\right] = 1 - e^{-0.1t}\left[\cos(0.995t) + 0.1005\sin(0.995t)\right]$$

Now

$$Z\left\{\mathcal{L}^{-1}\left[\frac{G(s)}{s}\right]\right\} = \frac{1}{1 - z^{-1}} - \frac{1 - 0.416z^{-1}}{1 - 0.985z^{-1} + 0.819z^{-2}}$$

for $T = 1$ second.   Multiplying by $(1 - z^{-1})$ and simplifying, we have

$$G(z) = \frac{0.431\,(z + 0.935)}{z^2 - 0.985z + 0.819}$$

The reader is asked to compare this result with the transfer function given in Example 12.5 and to provide a computer simulation with a unit step input. Is it performing as expected?

At this point we have all the information needed to look at the closed-loop transfer function in the discrete sense.

13.3 DISCRETE CLOSED-LOOP TRANSFER FUNCTION

If we refer to Fig. 5.1, the equivalent discrete control system can be generated. This is displayed in Fig. 13.4.

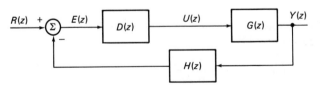

Figure 13.4   Discrete canonical form of control system

Following the derivation of Chapter 5.3, the discrete closed-loop transfer function is

$$P(z) = \frac{Y(z)}{R(z)} = \frac{D(z)G(z)}{1 + D(z)G(z)H(z)} \qquad (13.3.1)$$

with the error being

$$E(z) = \frac{R(z)}{1 + D(z)G(z)H(z)} \qquad (13.3.2)$$

The closed-loop discrete characteristic equation is

$$1 + D(z)G(z)H(z) = 0 \qquad (13.3.3)$$

EXAMPLE 13.4

Consider the discrete control system shown in Fig. 13.5. Calculate the following:

(a) Closed-loop transfer function $P(z)$.

(b) Steady-state error for a unit step input.

(c) Location of system pole on z-plane as gain varies from zero to infinity.

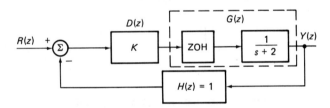

Figure 13.5   Control system for Example 13.4

SOLUTION

(a) To obtain $G(z)$ we have to apply ZOH. If we refer back to Example 13.1, this has already been evaluated. Using $G(z)$ from Example 13.1, equation (13.3.1), we have

$$P(z) = \frac{0.0906K}{z - 0.819 + 0.0906K}$$

(b) The error is given by equation (13.3.2), which is

$$E(z) = \frac{z - 0.819}{(1 - z^{-1})(z - 0.819 + 0.0906K)}$$

The steady-state error $e(\infty)$ is obtained by applying the final-value theorem. That is,

$$e(\infty) = \lim_{z \to 1} (1 - z^{-1}) \cdot E(z) = \frac{0.181}{0.181 + 0.0906K}$$

We find the steady-state error decreasing as K increases. This is the same result as in the continuous system.

(c) This in fact will produce a root locus for the discrete control system. The closed-loop characteristic equation is

$$z - 0.819 + 0.0906K = 0$$

Figure 13.6 provides the migration of the pole as the gain increases. It is interesting to note that our system becomes unstable when the gain is equal to 20. The reader should remember that this plant was stable for all values of gain in the continuous control system.

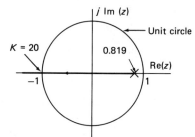

Figure 13.6   Migration of system pole for Example 13.4

The preceding example introduced the idea of the root locus in the z-plane. This process is very complicated for systems that are of an order greater than two. For this reason we will concentrate on first- and second-order systems. Design engineers will use the s-plane to design the required compensating network and then transfer the same into the z-plane by using the ZOH process. The brief encounter with the root locus on the z-plane will help the reader to understand the implications of sampling on systems which were considered stable at all times in the continuous domain.

13.4 ROOT LOCUS ON THE z-PLANE

If we consider first-order systems, we find the root locus following the path indicated by Fig. 13.6. The locus crosses the unit circle when $z = -1$. If we substitute this value into the closed-loop characteristic equation, we will obtain the maximum gain possible in the loop.   This gain will decrease if our sampling time increases.   This is the result of having the pole closer to the origin in the z-plane.   You were asked to predict this in problem 12.3.   How close were you?

The root locus for a second-order plant is more demanding.   Before we present the procedure, we have to investigate some terms associated with continuous second-order plants or systems.

Consider the following characteristic equation:

$$s^2 + 2\zeta_n \omega_n s + \omega_n^2 = 0 \tag{13.4.1}$$

If the roots are imaginary, we have

$$s_{1,2} = -\zeta_n \omega_n \pm j\omega_n \sqrt{1 - \zeta_n^2} \tag{13.4.2}$$

which are mapped into the z-plane by

$$z_{1,2} = e^{s_{1,2}T} \tag{13.4.3}$$

yielding

$$z_{1,2} = e^{-\zeta_n \omega_n T \pm j\omega_n \sqrt{1-\zeta_n^2}T} \tag{13.4.4}$$

If we consider one pole, we have

$$z_1 = e^{-\zeta_n \omega_n T} \left(\cos \left[\omega_n \sqrt{1 - \zeta_n^2}\, T \right] + j \sin \left[\omega_n \sqrt{1 - \zeta_n^2}\, T \right] \right)$$

Figure 13.7 provides the location of this pole on the z-plane.

From Fig. 13.7 we have

$$R = e^{-\zeta_n \omega_n T} \tag{13.4.6}$$

and

$$\cos \theta = \cos \left(\omega_n \sqrt{1 - \zeta_n^2}\, T \right) \tag{13.4.7}$$

which reduces to

$$\theta = \omega_n \sqrt{1 - \zeta_n^2}\, T \tag{13.4.8}$$

If we substitute equation (13.4.8) into (13.4.6), we have

$$R = e^{-\zeta_n \theta / \sqrt{1 - \zeta_n^2}} \tag{13.4.9}$$

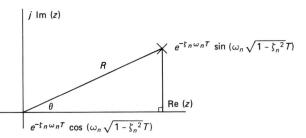

Figure 13.7   Pole location on z-plane

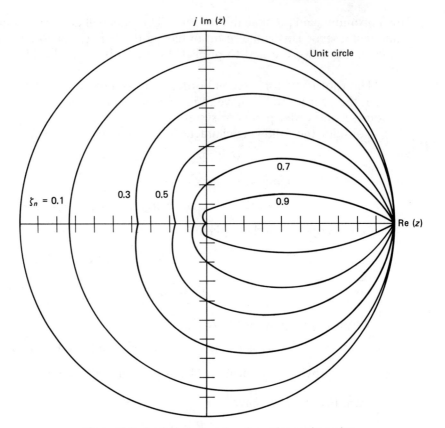

Figure 13.8   Loci of constant damping ratios on the z-plane

Equation (13.4.9) represents the loci for constant damping ratio on the z-plane. This will help in the design of a discrete control system to have the same characteristics as a continuous system.   Figure 13.8 provides the loci for various damping ratios.

If we are to use the damping ratio as a specification, we have to incorporate the discrete closed-loop characteristic equation.   The general second-order discrete closed-loop characteristic equation will be in the form

$$z^2 - bz + c = 0 \tag{13.4.10}$$

If the roots are imaginary, we have

$$z_{1,2} = \frac{b}{2} \pm j\,\frac{\sqrt{4c - b^2}}{2} \tag{13.4.11}$$

Figure 13.9 provides the location of one pole on the z-plane.

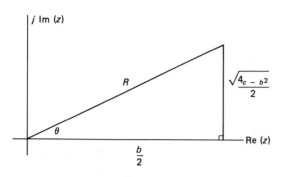

Figure 13.9 Pole location on z-plane

From Fig. 13.9 we have

$$\cos\theta = \frac{b}{2\sqrt{c}} \tag{13.4.12}$$

or

$$\theta = \cos^{-1}\left(\frac{b}{2\sqrt{c}}\right) \tag{13.4.13}$$

and

$$R = \sqrt{c} \tag{13.4.14}$$

The pole in Fig. 13.9 is the same pole as indicated in Fig. 13.7. Therefore equations (13.4.9) and (13.4.14) are equivalent. That is,

$$\sqrt{c} = e^{-\zeta_n\theta/\sqrt{1-\zeta_n^2}} \tag{13.4.15}$$

Substituting equation (13.4.13) into (13.4.15) and simplifying, we have

$$\cos\left[-\frac{\sqrt{1-\zeta_n^2}}{2\zeta_n}\ln c\right] - \frac{b}{2\sqrt{c}} = 0 \tag{13.4.16}$$

Equation (13.4.16) provides a means of calculating the gain required in a discrete control system so that it exhibits a given damping ratio. The equation is solved by iteration and is considerably more difficult as compared to the process in the continuous domain. Now we can continue with the root locus.

The root locus should be generated on the unit circle as indicated in Fig. 13.8. The reader is advised to reproduce Fig. 13.8 and use it as a basis for each root locus. Let's investigate the steps involved in generating a root locus on the z-plane for a second-order plant. We will assume the plant poles do not have imaginary locations.

Step 1.   Use the open-loop discrete transfer function $[D(z)G(z)H(z)]$ to plot the location of the poles and zeroes on the z-plane. This transfer function has to remain second order. Using $z = 1$ as a reference, the root locus occupies the real axis which is to the left of an odd number of poles and zeroes. This should sound familiar from the Laplace domain. The idea of breakaway and breakin points is also similar.

Step 2. If a breakaway point exists, then the locus is a circle with center at the zero and radius equal to the distance from the zero to the breakaway point. The breakaway point can be calculated from the characteristic equation by setting the discriminant equal to zero. That is,

$$b^2 - 4c = 0 \qquad (13.4.17)$$

This will generate a value for the gain. The solution of the characteristic equation with this gain will yield the breakaway point.

Step 3. We would like to know the maximum gain possible in our system and also the gain required to exhibit a specified damping ratio. If the locus leaves the unit circle at $z = \pm 1$, then the maximum gain is found by substituting the same back into the characteristic equation. If the locus leaves at any other point on the unit circle, we know that the radius R is unity. From equation (13.4.14) we have

$$c = 1 \qquad (13.4.18)$$

If the locus crosses the damping ratio curve specified, then equation (13.4.16) can be used to solve for the required gain via iteration.

It should be remembered that open-loop poles tend to open-loop zeroes as the gain approaches infinity. This is similar to the Laplace domain. Let's investigate with an example.

EXAMPLE 13.5

The second-order plant of Example 13.2 is to be proportionally controlled via computer. The system is to operate with a damping ratio of 0.707. What is the required gain? What is the maximum gain possible in the system? Provide an accurate root locus on the z-domain. Assume the system has unity feedback.

SOLUTION   From Example 13.2 we obtain $G(z)$, $D(z) = K$, and $H(z) = 1$. Therefore the open-loop transfer function is

$$D(z)G(z)H(z) = K \frac{0.316z + 0.0849}{z^2 - 0.418z + 0.0183}$$

The zero is located at $z = -0.269$. The poles are located at $z = 0.368$

and $z = 0.0497$. The characteristic equation is

$$z^2 - z\,(0.418 - 0.316K) + 0.0183 + 0.0849K = 0$$

The discriminant is given by

$$(0.418 - 0.316K)^2 - 4(0.0183 + 0.0849K)$$

If the discriminant is to be zero, we have $K = 0.17$ and $K = 5.86$. If we substitute these values back into the characteristic equation, we obtain $z = 0.18$ and $z = -0.72$. These represent the breakaway and breakin points, respectively. This implies that our locus leaves the unit circle at $z = -1$. If we substitute this value for z in the characteristic equation, we obtain a maximum gain of 6.24.

If we require a damping ratio of 0.707, equation (13.4.16) becomes

$$\cos\left[-0.5\ln\left(0.0183 + 0.0849K\right)\right] - \frac{0.418 - 0.316\,K}{2\,\sqrt{0.0183 + 0.0849\,K}} = 0$$

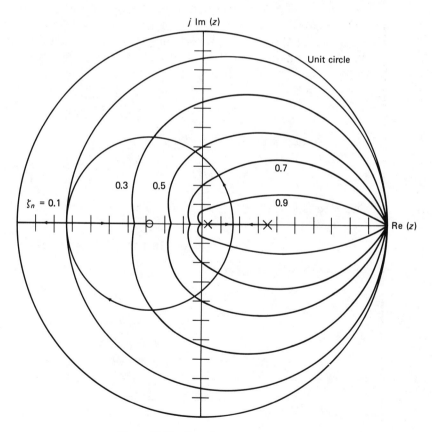

Figure 13.10   Root locus for Example 13.5

and by iteration generates a gain of 0.75. Figure 13.10 provides the accurate root locus on the z-plane. The reader is asked to find the gain required for a damping ratio of 0.707 if this system was a continuous one. Why the difference? Think about sampling time and the idea of ZOH.

Now let's look at another root-locus example.

EXAMPLE 13.6

Consider the control system as given in Fig. 13.11. Provide an accurate root locus for the system. What is the required gain for a damping ratio of 0.707? What is the maximum gain allowed? The selection of sampling time is variable.

SOLUTION It is evident that we have the same plant and for a good reason. You should have found the continuous control system requiring a larger gain to attain a 0.707 damping ratio in the preceding example. The ZOH process maintains a higher control action for a longer time, as would a continuous compensator. This implies a surplus in gain. Let's sample our system at 0.1 second and see if our digital gain approaches the continuous gain.

It is left as an exercise to show that at $T = 0.1$ second, the ZOH process yields

$$G(z) = \frac{0.00885z + 0.0076}{z^2 - 1.646z + 0.670}$$

Therefore our open-loop transfer function is

$$D(z)G(z)H(z) = K\frac{0.00885z + 0.0076}{z^2 - 1.646z + 0.670}$$

The zero is located at $z = -0.86$. The poles are located at $z = 0.74$ and $z = 0.91$. The characteristic equation is

$$z^2 - z(1.646 - 0.00885K) + 0.670 + 0.0076K = 0$$

The discriminant is given by

$$(1.646 - 0.00885K)^2 - 4(0.67 + 0.0076K)$$

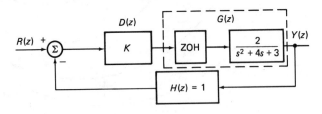

Figure 13.11 Control system for Example 13.10

If the discriminant is to be zero, we have $K = 0.49$ and $K = 759.2$. If we substitute these values back into our characteristic equation, we obtain $z = 0.82$ and $z = -2.54$. These represent the breakaway and breakin points, respectively. This implies our locus leaves the unit circle at some point other than $z = \pm 1$. Therefore we can use equation (13.4.18) to find the maximum gain. The maximum gain turns out to be 43.4.

If we require a damping ratio of 0.707, equation (13.4.16) becomes

$$\cos\left[-0.5 \ln (0.67 + 0.0076K)\right] - \frac{1.646 - 0.00885K}{2\sqrt{0.67 + 0.0076K}} = 0$$

and by iteration generates a gain of 2.1. Figure 13.12 provides the accurate root locus on the z-plane. The continuous system requires a gain of 2.5 to obtain the same damping ratio. The digital system increases the loop gain as sampling time increases. This is the result of the ZOH process.

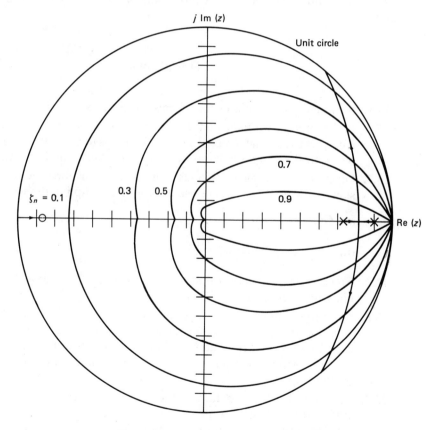

Figure 13.12   Root locus for Example 13.6

13.5 SUMMARY

In this chapter we introduced the ZOH process. We found that it provided the means to transform our continuous transfer function into the z-domain. The discrete closed-loop transfer function was generated and the analysis of a proportionally controlled first-order system was investigated. It was found that this system could be unstable in the discrete control scheme but would be stable at all times in the continuous system.

The root locus of a second-order system was investigated. The damping ratio from the continuous system was incorporated in our discrete domain. This provided a logical avenue for comparison with the continuous form of control. The root locus for a system with different sampling times was generated. It was found that the ZOH process provided excess gain in the loop. This effect can be minimized by increasing the sampling frequency or decreasing the digital gain. The former is more satisfactory, as it increases the system gain margin.

Problems

13.1. It was indicated that a D/A performs a ZOH process. What about an A/D?

13.2. In Chapter 12 we indicated the selection of the proper sampling time. Is it correct to say that a digital closed loop tends to instability as gain increases owing to the increased bandwidth with constant sampling time?

13.3. Given the following continuous transfer functions, provide the discrete versions using ZOH. Choose the appropriate sampling times.

(a) $G(s) = \dfrac{2}{s + 6}$

(b) $G(s) = \dfrac{s + 2}{s^2 + s + 2}$

13.4. Consider the DC motor outlined in Problem 6.5. Choose a suitable sampling time and generate $G(z)$ using ZOH.

13.5. Refer to Example 13.2. Generate $G(z)$ using the ZOH process and leaving T as a variable. Check your transfer function at $T = 1$ second. If it is correct, provide a computer simulation for the transfer function relative to a unit step input. Run the program for various values of T. Comment on the response.

13.6. The plant in Problem 13.4 is put into a proportionally controlled system with unity feedback. Calculate
(a) The closed-loop transfer function $P(z)$.
(b) Steady-state error for a unit step input. Assume $K = 0.5$.

13.7. For the control system in Problem 13.6,
  (a) Provide an accurate root locus on the z-plane.
  (b) Calculate the maximum gain.
  (c) Calculate the gain if the damping ratio is to be 0.707 and compare to result obtained in Problem 6.5.

13.8. Given the following plant. If $T = 0.3$, damping ratio is 0.3. Show that the gain required in a unity feedback system is 2.5.

$$G(s) = \frac{10}{s^2 + 7s + 10}$$

13.9. Given the following plant. Provide an accurate root locus if it is to be proportionally controlled with unity feedback. Choose the appropriate sampling time in applying the ZOH process. What is the gain required for a 0.707 damping ratio?

$$G(s) = \frac{s}{s^2 + 3s + 2}$$

13.10. Repeat Problem 13.9 for the following plant. How does the root locus compare with your prediction in Problem 12.5?

$$G(s) = \frac{s - 2}{s^2 + 3s + 2}$$

13.11. Show that the damped natural frequency for the second-order digital system is given by

$$\omega_d = \frac{1}{T} \cos^{-1}\left(\frac{b}{2\sqrt{c}}\right)$$

How does this compare with equation (11.4.5)?

13.12. Show that the settling time for the second-order digital system is given by

$$T_s = -\frac{5T}{\ln \sqrt{c}}$$

13.13. Show that the damping ratio for the second-order digital system is given by

$$\zeta_n = \left[\left(\frac{2\cos^{-1}\left(\frac{b}{2\sqrt{c}}\right)}{\ln c}\right)^2 + 1\right]^{-1/2}$$

13.14. The following is the characteristic equation for a digitally controlled second-order system. If the system is to have a damped natural frequency of 2.89

rad/sec, what is the required digital gain? The system is being sampled at 0.25 second.

$$z^2 - z(0.861 - 0.025K) + 0.176 + 0.0204K = 0$$

13.15. The following is the closed-loop transfer function for a digital control system. If the system has a natural undamped frequency (ω_n) of 7.62 rad/sec and a damping ratio of 0.5, what is the sampling time?

$$P(z) = \frac{0.198\,(z + 1)}{z^2 - 1.242z + 0.242}$$

13.16. The following is the closed-loop transfer function for a digital control system. What is the system damping ratio?

$$P(z) = \frac{0.00197\,(z + 1)}{z^2 - 1.922z + 0.9262}$$

13.17. Generate a computer program (language of your choice) to provide a root locus on the z-plane for a second-order control system. The program should graphically show the pole migration in the unit circle. It should also calculate the maximum gain possible and the gain for a specified damping ratio. The program can be tested using the examples in this chapter. The display should take the form of Fig. 13.8.

13.18. Consider the following plant. Provide an analog simulation for the plant. Build the circuit. Close the loop with a computer, as indicated in Fig. P13.18. Generate a computer program to proportionally control the plant. Use a setpoint of unity and your choice of sampling time. Monitor the plant output with various sampling times. How does it compare with theoretical predictions?

$$G(s) = \frac{1}{s^2 + 2.5s + 1}$$

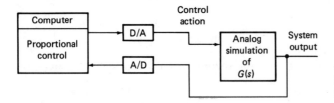

Figure P13.18 Computer control for Example 13.18

Part II
Discrete

14

Discrete Proportional Integral Derivative (P.I.D.) Control

14.1 INTRODUCTION

In Chapter 7 we investigated the theory of continuous P.I.D. control. All the details discussed there still apply, and if by chance you have forgotten anything, it is recommended that you reread that chapter. In this chapter we will develop a computer algorithm to perform the same type of control, and, since the computer is sampling, we designate our control to be discrete. The idea of sampling implies that our control system can exhibit instability with improper gains and sampling rates. The algorithm will be presented in the three most common computer languages: BASIC, C, and Pascal. It is assumed that the algorithm will be run on a real-time operating system with multitasking capabilities in order to perform multiple P.I.D. control loops.

14.2 DERIVATION OF THE ALGORITHM

If we consider the time-domain equivalent of P.I.D. control, we have

$$u(t) = K_p e(t) + K_i \int e(t)\, dt + K_d \frac{d}{dt} e(t) \qquad (14.2.1)$$

From (14.2.1) it is evident that our algorithm has to approximate the integral and derivative portion of the compensator. We will use the trapezoidal approximation for the former and a backward difference equation for the latter. As always, we will start with the easier proportional content. If $u_p(k)$ represents the present control action and e_k the present error, we have

$$u_p(k) = K_p \cdot e_k \qquad (14.2.2)$$

Owing to the proportional content, and, as in Chapter 7, K_p represents the proportional gain. Figure 14.1 provides the graphical implications of this portion of the algorithm.

To obtain the integral portion we have to find the area under the error curve from time zero to present sample time (kT). This provides some insight as to why this portion of control is referred to as past tense. From Fig. 14.2 the area in question is

$$A_0 + A_1 + \cdots + A_{k-2} + A_{k-1} \qquad (14.2.3)$$

which reduces to

$$\tfrac{1}{2}T[(e_0 + e_1) + (e_1 + e_2) + \cdots + (e_{k-2} + e_{k-1}) + (e_{k-1} + e_k)] \qquad (14.2.4)$$

and, multiplying this by our integral gain (K_i), we generate the integral portion of our P.I.D. compensator $u_i(k)$:

$$u_i(k) = \tfrac{1}{2}K_i T[(e_0 + e_1) + (e_1 + e_2) + \cdots + (e_{k-2} + e_{k-1}) + (e_{k-1} + e_k)] \qquad (14.2.5)$$

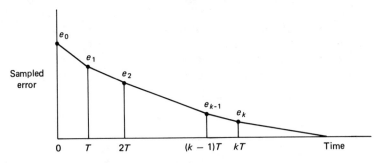

Figure 14.1 Discrete proportional control

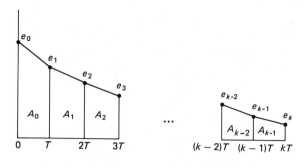

Figure 14.2 Trapezoidal approximation to area

To find the derivative portion we simply find the rate of change of error at time kT. This can be found by calculating the slope of the line between our present error (e_k) and our previous error (e_{k-1}). Figure 14.3 provides the graphical implications.

From Fig. 14.3 the derivative at time kT is

$$\frac{1}{T}(e_k - e_{k-1}) \tag{14.2.6}$$

and, multiplying this by our derivative gain K_d, we generate the derivative portion of our P.I.D. compensator $u_d(k)$:

$$u_d(k) = \frac{K_d}{T}(e_k - e_{k-1}) \tag{14.2.7}$$

To generate our P.I.D. control action $u(k)$, we sum the individual components as follows:

$$u(k) = u_p(k) + u_i(k) + u_d(k) \tag{14.2.8}$$

and by substitution we get

$$u(k) = K_p e_k$$

$$+ \frac{1}{2} K_i T[(e_0 + e_1) + (e_1 + e_2) + \cdots + (e_{k-2} + e_{k-1}) + (e_{k-1} + e_k)]$$

$$+ \frac{K_d}{T}(e_k - e_{k-1})$$

$$\tag{14.2.9}$$

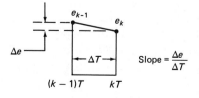

Slope $= \dfrac{\Delta e}{\Delta T}$

Figure 14.3 Derivative via backward difference equation

If we inspect (14.2.9), it is evident that our control action is not in a form usable by a computer. Let's look at the control action one sample ago. This can be obtained by replacing the subscript k with $k - 1$ in (14.2.9). Doing this, we have

$$u(k - 1) = K_p e_{k-1}$$

$$+ \frac{1}{2} K_i T[(e_0 + e_1) + (e_1 + e_2) + \cdots + (e_{k-2} + e_{k-1})] \qquad (14.2.10)$$

$$+ \frac{K_d}{T}(e_{k-1} - e_{k-2})$$

and subtracting (14.2.10) from (14.2.9) yields

$$u_k - u_{k-1} = K_p(e_k - e_{k-1})$$

$$+ \frac{K_i T}{2}(e_k + e_{k-1}) \qquad (14.2.11)$$

$$+ \frac{K_d}{T}(e_k - 2e_{k-1} + e_{k-2})$$

which simplifies to

$$u_k - u_{k-1} = \alpha e_k + \beta e_{k-1} + \gamma e_{k-2} \qquad (14.2.12)$$

where

$$\alpha = \left(K_p + \frac{K_i T}{2} + \frac{K_d}{T} \right) \qquad (14.2.13)$$

$$\beta = \left(\frac{K_i T}{2} - K_p - \frac{2K_d}{T} \right) \qquad (14.2.14)$$

$$\gamma = \left(\frac{K_d}{T} \right) \qquad (14.2.15)$$

If we rewrite (14.2.12), we have

$$\boxed{u_k = u_{k-1} + \alpha e_k + \beta e_{k-1} + \gamma e_{k-2}} \qquad (14.2.16)$$

From (14.2.16) it is evident that this equation is quite easily implemented on a computer. The present control action is simply the sum of the previous control action, present error, and previous two errors multiplied by the proper constants alpha, beta, and gamma. It is also interesting to note that once a computer program is created, we can perform any combination of control covered in Chapter 7 by simply setting the appropriate gain to zero.

Now let's have a look at the program.

14.3 COMPUTER IMPLEMENTATION OF P.I.D. ALGORITHM

BASIC Language

Figure 14.4 provides a sample program for P.I.D. control utilizing the BASIC language.

```
10   rem *** BASIC version of P.I.D. algorithm *** V J B ***
20   rem
30   rem ** shift error and control action stacks ***
40   rem
50   for sample = 1 to 0 step -1
60       cntlaction(sample + 1) = cntlaction(sample)
70       syserror(sample + 1)   = syserror(sample)
80   next sample
90   rem *** read feedback via A/D and generate present error ***
100  rem
110  syserror(0) = setpoint - adin
120  rem *** calculate new control action and put in control action ***
130  rem *** stack.
140  rem
150  cntlaction(0) = cntlaction(1) +alpha*syserror(0) +beta*syserror(1)
160                 +gamma*syserror(2)
170  rem
180  rem *** output control action to system via D/A ***
190  rem
200  daout = cntlaction(0)
210  rem
220  rem *** end of algorithm ***
230  end
```

Figure 14.4 BASIC program for P.I.D. algorithm

C Language

Figure 14.5 provides a sample program for P.I.D. control utilizing the C language.

```
#include <stdio.h>
#include <math.h>
main()          /* C version of P.I.D. algorithm - V J B --*/
{

        int sample,setpoint;
        int adin,daout;
        static float alpha,beta,gamma;
        static float cntlaction[3],syserror[3];
        for (sample = 1; sample >=0; sample--) {

                /* make room for present control action and error*/
                /* on appropriate stack.*/

                cntlaction[sample+1] = cntlaction[sample];
                syserror[sample+1] = syserror[sample];
                }

        /* read feedback via A/D and generate present error*/

        syserror[0] = setpoint - adin;

        /*calculate new control action and put in control action stack*/

        cntlaction[0] = cntlaction[1]
                        +alpha*syserror[0]
                        +beta*syserror[1]
                        +gamma*syserror[2];

                /*output control action to system via D/A*/

        daout = cntlaction[0];

        /*end of algorithm*/
```

Figure 14.5 C program for P.I.D. algorithm

Pascal Language

Figure 14.6 provides a sample program for P.I.D. control utilizing the Pascal language.

```pascal
program pid_control;

{PASCAL version of P.I.D. algorithm ** V J B **)

var

sample,setpoint : integer;
adin,daout      : integer;
alpha,beta,gamma: real;
cntlaction      : array[0..1] of real;
syserror        : array[0..1] of real;

{shift error and control action stacks}

begin

    for sample := 1 downto 0 do
    begin

        cntlaction[sample + 1] := cntlaction[sample];
        syserror[sample + 1]   := syserror[sample];

    end;

{read feedback via A/D and generate present error}

syserror := setpoint - adin;

{calculate new control action and put in control action stack}

cntlaction[0] := cntlaction[1]
                +alpha*syserror[0]
                +beta*syserror[1]
                +gamma*syserror[2];

{output control action to system via D/A}

daout := cntlaction[0];

{end of algorithm}

end.
```

Figure 14.6   Pascal program for P.I.D. algorithm

Machine-Independent Algorithm Structure

Notice that the programs above offer a sample P.I.D. algorithm and will change to reflect the syntax requirements of your machine. The real-time element has been omitted, and it should be remembered that the programs will have to be executed at intervals decided upon by the designer to insure system stability. This sampling time required will also decide the number of P.I.D. loops your machine can handle and in some instances could be limited by the number of tasks that can be serviced by a multitasking system. Figure 14.7 provides a machine-independent structure of the P.I.D. algorithm at a register level.

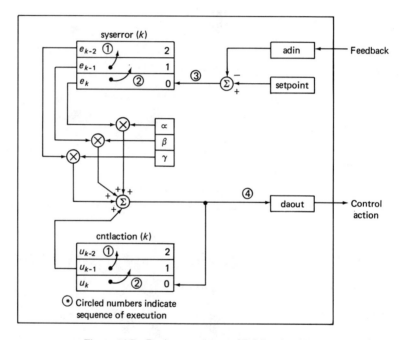

Figure 14.7   Register structure of P.I.D. algorithm

To verify our algorithm and its implications for system response we will run a computer simulation on the tank problem in Chapter 7. The gains will be the same as the optimized system in Example 7.4 so that a valid comparison can be made.

EXAMPLE 14.1

The gains and plant transfer function from Example 7.4 are

$$K_p = 9.5$$

$$K_i = 15.36$$

$$K_d = 0.25$$

$$G(s) = \frac{2}{s^2 + 7s + 6}$$

Using the zero-order hold approximation with a sampling time T of 0.1 second, we obtain

$$G(z) = \frac{H_2(z)}{U(z)} = (1 - z^{-1})\left\{ Z\left(\mathcal{L}^{-1}\left[\frac{G(s)}{s} \right] \right) \right\}_{T=0.1}$$

Simplifying, we get

$$\frac{H_z(z)}{U(z)} = \frac{0.008z^{-1} + 0.00632z^{-2}}{1 - 1.454z^{-1} + 0.4966z^{-2}}$$

and, taking the inverse z-transform, we have the difference equation

$$h_2(k) = 1.454h_2(k - 1) - 0.4966h_2(k - 2) + 0.008u_{k-1} + 0.00632u_{k-2}$$

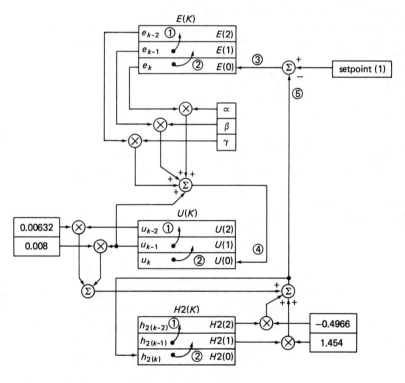

$\odot$ Circled numbers indicate
sequence of execution

Figure 14.8   Register structure of P.I.D. control simulation

In order to perform our simulation a computer program will be written to perform the modified version of Fig. 14.7 given in Fig. 14.8. From Fig. 14.8 it is evident that we are simulating our plant via computer and therefore require no contact with the outside world. The system output and control action can conveniently be recorded and graphed.

From Fig. 14.8 the BASIC program in Fig. 14.9 is created, and the system output and control action are given by $H2(0)$ and $U(0)$, respectively.

```
10   REM  *** P.I.D. control simulation of Example 14.1 ***
20   CLS
30   KEY OFF
40   INPUT "Proportional gain Kp = ",KP
50   INPUT "Integral Gain Ki = ",KI
60   INPUT "Derivative Gain Kd = ",KD
62   ALPHA = KP + KI*.1/2 + KD/.1
64   BETA = KI*.1/2 -KP - 2*KD/.1
66   GAMMA = KD/.1
70   REM  ** Take 100 samples at 0.1 sec/sample **
80   FOR SAMPLE = 0 TO 99
90       FOR K = 1 TO 0 STEP -1
100          E(K+1) = E(K)
110          U(K+1) = U(K)
120          H2(K+1)= H2(K)
130      NEXT K
140   E(0) = 1 - H2(0)
150   U(0) = U(1) + ALPHA*E(0) + BETA*E(1) + GAMMA*E(2)
160   H2(0) = 1.454*H2(1)-.4966*H2(2)+8.000001E-03*U(1)+.00632*U(2)
170   PRINT SAMPLE,U(0),H2(0)
180   NEXT SAMPLE
190   END
```

Figure 14.9 BASIC program for P.I.D. simulation

If we run the program in Fig. 14.9, we generate the control action and system output in Figs. 14.10 and 14.11, respectively. From Fig. 14.11 (page 278) we find the system having a very low damping ratio, and from Fig. 14.10 the control valve will have a momentary negative voltage applied. What has happened? In Example 7.4 we optimized our system in order to exhibit a

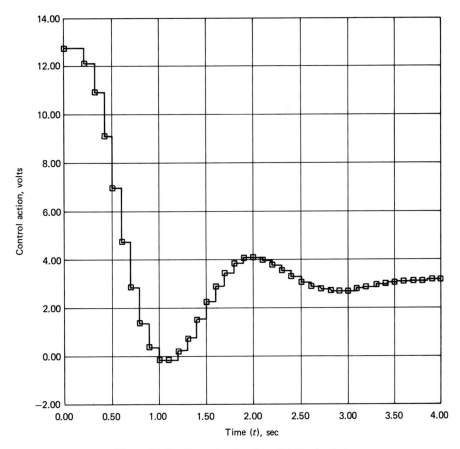

Figure 14.10   Control action from P.I.D. simulation

system damping ratio of 0.707 and a settling time of 3 seconds, and we now find that this is no longer true.

At this point the reader should load the program in Fig. 14.9 into a computer, make the relevant syntax corrections required for his machine, and run the program, varying the gains in such a fashion that the system response resembles that of Example 7.4.   Having completed this, the reader is asked to come up with an explanation for the gains required in this digital system. In the next section we will explore this irregularity in terms of system sampling time, and our program in Fig. 14.9 will be modified to allow the simulation at different sampling times.

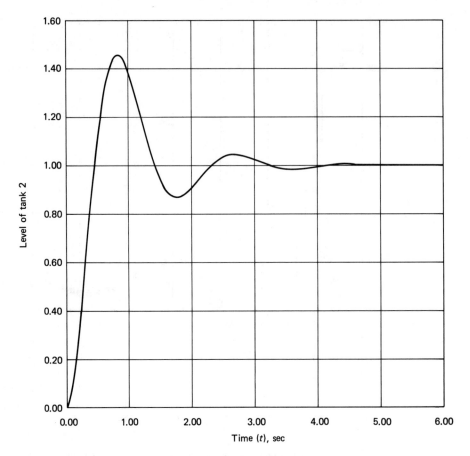

Figure 14.11　System output from P.I.D. simulation

14.4 EFFECT OF SAMPLING TIME ON P.I.D. ALGORITHM

From Fig. 14.11 it would appear that our system has a surplus of integral gain (K_i) even though we transformed our optimal system from the continuous to the discrete domain. If we look at the implications of this transformation, perhaps we can explain the surplus of integral gain. Equation (14.2.16) represents the difference equation for our discrete P.I.D. algorithm. If we take z-transforms, we have

$$U(z) = U(z)z^{-1} + \alpha E(z) + \beta E(z)z^{-1} + \gamma E(z)z^{-2} \qquad (14.4.1)$$

and the transfer function $D(z)$ is

$$D(z) = \frac{U(z)}{E(z)} = \frac{\alpha z^2 + \beta z + \gamma}{z(z-1)} \tag{14.4.2}$$

If this is to represent a continuous P.I.D. compensator, then our sampling time T should approach zero. Letting $T \to 0$, equations (14.2.13) through (14.2.15) would be

$$\alpha \to \frac{K_d}{T} \tag{14.4.3}$$

$$\beta \to -\frac{2K_d}{T} \tag{14.4.4}$$

$$\gamma = \frac{K_d}{T} \tag{14.4.5}$$

respectively. If we substitute (14.4.3)–(14.4.5) into (14.4.2), we have

$$D(z) \to \frac{\dfrac{K_d}{T} z^2 - \dfrac{2K_d}{T} z + \dfrac{K_d}{T}}{z(z-1)} \tag{14.4.6}$$

which simplifies to

$$D(z) \to \frac{\dfrac{K_d}{T}(z-1)(z-1)}{z(z-1)} \tag{14.4.7}$$

From (14.4.7) we find the zeroes of the compensating network approaching the unit circle. If we are to better approximate the continuous compensator, we should try to move the zeroes on the z-plane toward the unit circle. If our sampling time is constant, we see from equations (14.2.13)–(14.2.15) that our only alternative is to decrease both the proportional and integral gains so that equations (14.4.3)–(14.4.5) are approximated. This implies a surplus of both gains as compared to the optimal system designed in the continuous sense. It is important to note that the proportional and integral gain still affect the system response as indicated in Chapter 7.

One can conclude that an optimally designed P.I.D. compensator in the continuous sense can be readily transferred into the discrete domain. It should be understood that we can attain analogous responses if our sampling time is very small; if it is not small enough, then a reduction in proportional and integral gain is required. The example that follows will help to clarify these points.

EXAMPLE 14.2

We will run a simulation of our tank system in Chapter 7 with the sampling time left as a variable. The program in Fig. 14.9 will be modified to allow these changes and run with the following constraints:

(a) Sampling time T is constant at 0.1 sec with the proportional and integral gain less than that calculated in Example 7.4.

(b) Sampling time T is decreased from 0.1 sec with the proportional and integral gain the same as that calculated in Example 7.4.

To modify the program in Fig. 14.9 we must find the system difference equation in terms of the sampling time T. Using zero-order hold with a sampling time of T seconds on the plant transfer function ($G(s)$ from Example 7.4), we have, after simplification, the discrete plant transfer function $G(z)$:

$$G(z) = \frac{Hz(z)}{U(z)} = \frac{\begin{aligned}(0.06667e^{-6T} - 0.4e^{-T} + 0.3333)z^{-1} \\ + (0.3333e^{-7T} + 0.06667e^{-T} - 0.4e^{-6T})z^{-2}\end{aligned}}{1 - (e^{-6T} + e^{-T})z^{-1} + e^{-7T}z^{-2}} \qquad (14.4.8)$$

and, taking the inverse z transform, we have the desired difference equation:

$$h_2(k) = (e^{-6T} + e^{-T})h_2(k - 1) - (e^{-7T})k_2(k - 2)$$

$$+ (0.06667e^{-6T} - 0.4e^{-T} + 0.3333)u_{k-1} \qquad (14.4.9)$$

$$+ (0.3333e^{-7T} + 0.06667e^{-T} - 0.4e^{-6T})u_{k-2}$$

From (14.4.9) we can modify the BASIC program given in Fig. 14.9 to that given in Fig. 14.12.

```
5    REM *** P.I.D. control simulation of Example 14.2 ***
10   CLS
20   KEY OFF
30   INPUT "proportional gain Kp = ",KP
40   INPUT "integral gain Ki = ",KI
50   INPUT "derivative gain Kd = ",KD
55   INPUT "sampling time T = ",T
60   ALPHA = KP +KI*T/2 + KD/T
70   BETA = KI*T/2 - KP - 2*KD/T
80   GAMMA = KD/T
90   REM ** sample for 10 seconds **
110  FOR SAMPLE = 0 TO (10/T-1)
120      FOR K = 1 TO 0 STEP -1
130          E(K+1) = E(K)
```

Figure 14.12 BASIC program for P.I.D. simulation

```
140            U(K+1) = U(K)
150            Y(K+1) = Y(K)
160      NEXT K
170  E(0) = 1 - Y(0)
180  U(0) = U(1)+ALPHA*E(0)+BETA*E(1)+GAMMA*E(2)
190  Y(0)=(EXP(-6*T)+EXP(-T)*Y(1)-EXP(-7*T)*Y(2)+(.06667*EXP(-6*T)
            -.4*EXP(-T)+.3333)*U(1)+(.3333*EXP(-7*T)+.06667*EXP(-T)
            -.4*EXP(-6*T))*U(2)
210  PRINT SAMPLE*T,U(0),Y(0)
220  NEXT SAMPLE
240  END
```

Figure 14.12   (con't.)

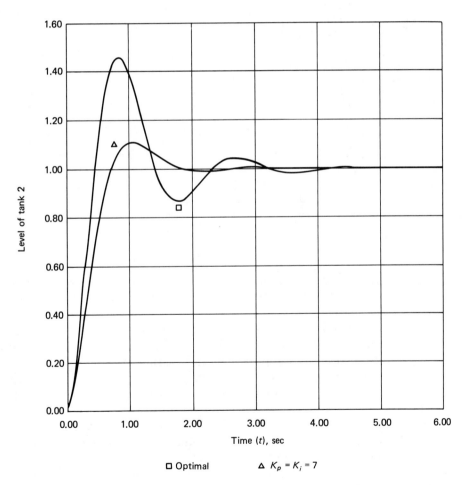

□ Optimal   △ $K_p = K_i = 7$

Figure 14.13   Comparison of system response with constant sampling time ($T =$ 0.1 sec) and decreasing proportional integral gain

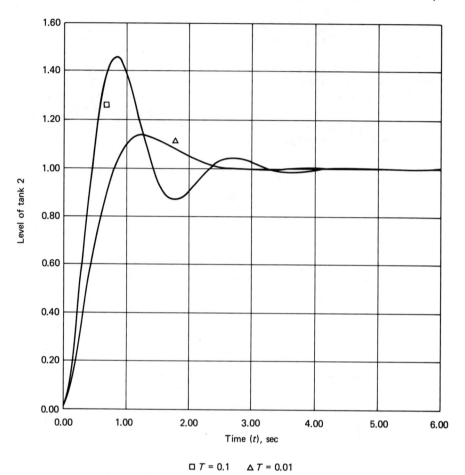

Figure 14.14 Comparison of system output with constant proportional integral gain and changing sampling time

If we run the program in Fig. 14.12 with the constraints given in (a) and (b), we have the responses given in Figs. 14.13 and 14.14, respectively. As expected, the responses approximate the continuous system given in Fig. 7.25. The reader is encouraged to load this program in his computer and experiment with changes in both sampling times and individual gains. The reader should plot the responses and convince himself that the gain-to-sampling-time tradeoff is in fact beneficial for the proper application of the discrete P.I.D. algorithm. This tradeoff becomes important in a multitasking environment where multiple P.I.D. loops are being executed. If the system is running at near maximum CPU time, it is possible to save valuable execution time with this gain-to-sampling-time adjustment scheme. Naturally this can only be feasible if the increase in sampling time retains stability.

14.5 SUMMARY

In this chapter we developed the discrete version of the P.I.D.-type algorithm. The integral content was approximated by the trapezoidal rule and the derivative via backward difference equation. The algorithm was presented in the three most common computer languages, BASIC, C, and Pascal. It was presented in register form to provide a language-independent structure. The algorithm was tested on the system outlined in Example 7.4 and found to contain too much proportional integral gain. The abundance of gains could be removed by a smaller sampling time or reduction of the gains. This was verified via computer simulation, and the reader was asked to pursue this area by running various gains and sampling times with the simulation program provided.

Problems

14.1. From equation (14.2.12), find $D(z)$. Where are the poles located on the z-plane? How many zeroes are there? How does this compare with the continuous P.I.D. compensation network of Chapter 7?

14.2. From equation (14.2.12), find $D(z)$ for a P.I. controller. Show that the zero location is given by $z = -\beta/\alpha$.

14.3. Consider the P.I. control system shown in Fig. P14.3. The system has a sampling time of 0.1 second and is to exhibit a 0.707 damping ratio. The settling time is to be 2 seconds. Calculate the required proportional and integral gains. Problem 13.12 and equation (13.4.16) will help.

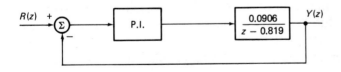

Figure P14.3 Control system for Problem 14.3

14.4. Provide a computer simulation for the control system designed in Problem 14.3. Has it met the specifications? Why? Try the following modifications to the plant difference equation:

$$u_k = 0.819u_{k-1} + 0.090e_k$$

Run the program and comment on the results.

14.5. Consider the P.I. control system given in Fig. P14.5. If $T = 1$ second, find the maximum values for the proportional and integral gains. [*Hint:* Consider marginal stability.]

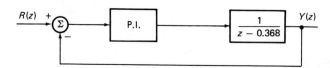

Figure P14.5 Control system for Problem 14.5

14.6. Generate the P.I.D. algorithm using rectangular approximation for area. Modify the program in Fig. P14.5 and run it. How does this algorithm compare with the original?

14.7. In Chapter 7 we investigated each segment of P.I.D. control separately. It is now possible to verify our theoretical predictions via computer. If you run the program in Fig. 14.12 leaving only one segment of the PID control active, then the outputs should reflect those of Chapter 7. That is,
(a) K_p = value, $K_i = K_d = 0$.
(b) K_i = value, $K_p = K_d = 0$.
(c) K_d = value, $K_p = K_i = 0$.
The reader may want to modify the program so that a graph is generated on the screen. Are the results as expected?

14.8. Generate a program to implement the P.I.D. algorithm on any favorite microprocessor. Calculate the execution time. How many P.I.D. loops could be handled if each loop required a sampling time of 0.5 second?

14.9. A common problem with a digital integrator is saturation. Consider a motor speed-control system using P.I. control. If you stall the motor for a period of time, the digital integration will saturate and keep the motor turning at full speed after you let go. Why? How can it be remedied?

14.10. Refer to the program in Fig. 14.12.   Add the following line:

```
185   IF SAMPLE < 1/T THEN 210
```

This provides a simulation of the transducer hesitating for 1 second at the beginning. Run the program for various gains and sampling times. Comment on the system response.

14.11. Refer to the program in Fig. 14.12. Add the following lines.

```
57    INPUT "Noise level = ",NOISE
185   U(0) = U(0) + NOISE*E(0)
```

This represents a noise which is proportional to the error signal and is added to the control action. Run the program for NOISE varying from −2 to 2. Use various gains and sampling times. Has the output level changed?

14.12. Refer to the program in Fig. 14.12.   Add the following lines.

```
57    INPUT "Offset = ",OFFSET
185   U(0) = U(0) + OFFSET
```

This represents an offset in the control action.   Run the program with various gains and sampling times.   Has the output level changed?

14.13. Refer to the program in Fig. 14.12.   Add the following lines.

```
57    INPUT "Offset = ", OFFSET
185   REM
200   Y(0) = Y(0) + OFFSET
```

This represents an offset in the feedback of our control system.   Run the program for various gains and sampling times.   Has the output level changed?

14.14. Consider the P.I.D. control system given in Fig. P14.14.   This represents the plant in Example 7.4 with $T = 0.1$ second.   If all gains are as in Example 7.4, find $D(z)$.   Find the open-loop transfer function.   Plot the poles and zeroes on the z-plane.   Indicate the regions on the real axis that would be included in a root locus for this system.   Show that the characteristic equation is given by

$$z^4 - 2.3519z^3 + 1.9214z^2 - 0.5632z + 0.01578 = 0$$

From the characteristic equation, plot the pole locations on the z-plane.   Are they following the pattern indicated by the root locus?   In Chapter 7 we placed the poles in a straight line.   In Chapter 12 we indicated that straight lines in the s-plane become circles in the z-plane.   The radius of the circle is given by

$$z = e^{sT}$$

where s represents the location in the s-plane.   In our system $s = -2.5$ and $T = 0.1$ seconds.   Draw this circle on the z-plane.   Are the poles on this circle?   Is this still an optimal system?   What effect was observed due to the ZOH process?

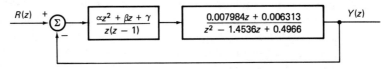

Figure P14.14   Control system for Problem 14.14

Part II
Discrete

15

Discrete Compensator Design

15.1 INTRODUCTION

In chapter 10 we provided the discrete version of a differential equation by approximating the derivative. In Chapter 13 we introduced the ZOH process and generated the discrete transfer function from a continuous one. Both of these methods provided a discrete version for continuous compensating networks. In this chapter we will investigate two additional techniques, which are borrowed from the area of digital filtering. The reader should appreciate that compensating networks are filters and that knowledge of one can be applied to the other. The *matched pole-zero* and *bilinear transformation* techniques will provide the transformation from the continuous to the discrete domain.

We will turn our attention to the lead, lag, and lead-lag compensators in the discrete domain. In many cases the design engineer will design the control system as outlined in Chapter 8 and transform his design using the techniques mentioned above. As the control engineer becomes comfortable in the z-domain, the design can be carried out using the methods of the root locus, as outlined in Chapter 13. We will investigate both processes. Let's begin with the matched pole-zero technique.

15.2 MATCHED POLE-ZERO TECHNIQUE

Consider the following transfer function:

$$D(s) = K \frac{(s + r_1)(s + r_2) \cdots (s + r_n)}{(s + p_1)(s + p_2) \cdots (s + p_m)} \qquad (15.2.1)$$

On applying the **matched pole-zero technique**, we have

$$D(z) = K_1 \frac{(z + 1)^{m-n}(z - e^{-r_1 T})(z - e^{-r_2 T}) \cdots (z - e^{-r_n T})}{(z - e^{-p_1 T})(z - e^{-p_2 T}) \cdots (z - e^{-p_m T})} \qquad (15.2.2)$$

On close inspection of equation (15.2.2), we find all the finite poles and zeroes of the continuous transfer function being transformed by $z = e^{sT}$. The zeroes at infinity are mapped to $z = -1$ and therefore a matched number of poles and zeroes. The new gain K_1 has to be calculated, such that the gain at the critical frequency of the original transfer function is matched.

EXAMPLE 15.1

Consider the following lowpass filter. Find the equivalent digital filter using the matched pole-zero technique.

$$D(s) = \frac{1}{s^2 + 0.2s + 1}$$

The reader should refer back to Example 13.3. This is the same transfer function and therefore we have a form of comparison. As in Example 13.3 we will use a sampling time of 1 second.

SOLUTION If we compare the transfer function with equation (15.2.1), we have $m = 2$ and $n = 0$. The pole locations are given by

$$p_1 = 0.1 - j0.995$$

and

$$p_2 = 0.1 + j0.995$$

Substituting this information into equation (15.2.2) and simplifying, we have

$$D(z) = K_1 \frac{(z + 1)^2}{z^2 - 0.985z + 0.819}$$

Since this transfer function represents a lowpass filter, the critical frequency is at DC. The original gain at this frequency is given by $s = 0$ and is equal to 1. If the digital filter is to have the same gain at DC ($z = 1$), then

$K_1 = 0.209.$ Therefore the digital filter is given by

$$D(z) = 0.209 \, \frac{z^2 + 2z + 1}{z^2 - 0.985z + 0.819}$$

The reader is asked to generate the frequency response for this digital filter and compare it to the response obtained in Example 12.5. This filter was also investigated in Problem 3.8 in the continuous sense. How does it compare?

Let's look now at the bilinear transformation.

15.3 THE BILINEAR TRANSFORMATION

The **bilinear transformation** technique involves direct substitution of the following into the continuous transfer function:

$$s = \frac{2}{T} \frac{z - 1}{z + 1} \tag{15.3.1}$$

Equation (15.3.1) evolves from the relationship

$$z = e^{sT} \tag{15.3.2}$$

from which

$$s = \frac{1}{T} \ln z \tag{15.3.3}$$

Now

$$\ln z = 2 \frac{z - 1}{z + 1} + \frac{1}{3} \left(\frac{z - 1}{z + 1} \right)^3 + \cdots \tag{15.3.4}$$

From (15.3.4) we find that the first term of the series provides a good approximation to the function; therefore, equation (15.3.1).

EXAMPLE 15.2

Find the digital filter for the continuous filter in Example 15.1. Use the bilinear transformation.

SOLUTION   The sampling time from Example 15.1 is 1 second. Therefore

$$D(z) = \frac{1}{4 \left(\dfrac{z - 1}{z + 1} \right)^2 + 0.4 \left(\dfrac{z - 1}{z + 1} \right) + 1}$$

and simplifying, we obtain

$$D(z) = \frac{0.185(z^2 + 2z + 1)}{z^2 - 1.111z + 0.852}$$

If we refer to Problem 13.11, we find the damped natural frequency of this digital filter to be 0.925 rad/sec. The original continuous filter had a damped natural frequency of 0.995 rad/sec. The difference is the result of the approximation to equation (15.3.4). If we could alter our original continuous transfer function by increasing the critical frequency, then the bilinear transformation would warp this frequency back to the required critical frequency.

From equation (15.2.1) every critical frequency has the form

$$T(s) = 1 + \frac{s}{\omega_n} \tag{15.3.5}$$

which has a frequency response of

$$T(j\omega) = 1 + j \tag{15.3.6}$$

at the critical frequency. If we prewarp the critical frequency of equation (15.3.5), we have

$$T_{PW}(s) = 1 + \frac{s}{\omega_{PW}} \tag{15.3.7}$$

Applying the bilinear transformation to equation (15.3.7), we obtain

$$T_{PW}(z) = 1 + \frac{2}{\omega_{PW}T}\frac{z - 1}{z + 1} \tag{15.3.8}$$

with the frequency response at the critical frequency being

$$T_{PW}(e^{j\omega_n T}) = 1 + \frac{2}{\omega_{PW}T}\frac{e^{j\omega_n T} - 1}{e^{j\omega_n T} + 1} \tag{15.3.9}$$

Now equations (15.3.9) and (15.3.6) are equal. Therefore

$$\frac{2}{\omega_{PW}T}\frac{e^{j\omega_n T} - 1}{e^{j\omega_n T} + 1} = j \tag{15.3.10}$$

Simplifying, we have

$$\frac{2}{\omega_{PW} \cdot T}[\cos(\omega_n T) + j\sin(\omega_n T) - 1] = -\sin(\omega_n T) + j[\cos(\omega_n T) + 1]$$

$$\tag{15.3.11}$$

From equation (15.3.11), we have

$$\frac{2}{\omega_{PW}T} [\cos (\omega_n T) - 1] = -\sin (\omega_n T) \tag{15.3.12}$$

$$\frac{2}{\omega_{PW}T} \sin (\omega_n T) = \cos (\omega_n T) + 1 \tag{15.3.13}$$

Substituting equation (15.3.12) into (15.3.13) and simplifying, we obtain

$$\left(\frac{\omega_{PW}T}{2}\right)^2 = \frac{1 - \cos (\omega_n T)}{1 + \cos (\omega_n T)} \tag{15.3.14}$$

Using double-angle trigonometric identities, we have

$$\left(\frac{\omega_{PW}T}{2}\right)^2 = \frac{\sin^2 \left(\dfrac{\omega_n T}{2}\right)}{\cos^2 \left(\dfrac{\omega_n T}{2}\right)} \tag{15.3.15}$$

and, simplifying, we obtain

$$\omega_{PW} = \frac{2}{T} \tan \left(\frac{\omega_n T}{2}\right) \tag{15.3.16}$$

The original critical frequency would be given by

$$\omega_n = \frac{2}{T} \tan^{-1} \left(\frac{\omega_{PW}T}{2}\right) \tag{15.3.17}$$

in terms of the prewarped frequency.
    Let's redo Example 15.2.

EXAMPLE 15.3

Apply prewarping to the lowpass filter in Example 15.1 so that the digital filter has the proper critical frequency.

SOLUTION   The standard form for the lowpass filter is

$$D(s) = \frac{\omega_n^2}{s^2 + 2\zeta_n \omega_n s + \omega_n^2}$$

and the prewarped form would be

$$D_{PW}(s) = \frac{\omega_{PW}^2}{s^2 + 2\zeta_n \omega_{PW} s + \omega_{PW}^2}$$

The original filter has $\omega_n = 1$ rad/sec, $\zeta_n = 0.1$. The sampling time is 1

second.   From equation (15.3.16), the prewarped frequency is 1.0926 rad/
sec.   Therefore the prewarped transfer function becomes

$$D_{PW}(s) = \frac{1.1938}{s^2 + 0.2185s + 1.1938}$$

And applying the bilinear transformation to this transfer function, we have

$$D(z) = \frac{0.212(z^2 + 2z + 1)}{z^2 - 0.9967z + 0.8448}$$

after considerable simplification.   The digital filter now has a damped natural
frequency of 0.997 rad/sec.   This is comparable to the continuous filter.

The reader is asked to generate the frequency response for both filters
attained by the bilinear transformation.   Compare all the frequency responses
and decide which method provides the most accurate filter.

15.4 DISCRETE LEAD COMPENSATION

In Chapter 8 we investigated the continuous discrete lead compensator.   We
found it to increase system speed and steady-state error.   The continuous form
was found to be

$$D(s) = K\frac{s + a}{s + b}; \qquad a < b \tag{15.4.1}$$

if we apply the matched pole-zero technique, we have

$$D(z) = K_1\frac{z - A}{z - B}; \qquad A > B \tag{15.4.2}$$

where

$$A = e^{-aT} \tag{15.4.3}$$

and

$$B = e^{-bT} \tag{15.4.4}$$

The governing difference equation for this compensator is

$$u_k = B \cdot u_{k-1} + K_1(e_k - A \cdot e_{k-1}) \tag{15.4.5}$$

The value for K_1 is left to be calculated in the design process.   Let's investigate
with an example.

EXAMPLE 15.4

A control engineer has finished the design of a lead compensator in the s-domain. It was appropriate to cancel the slow pole of the system with the compensating zero. Figure 15.1 provides the control strategy. This system is to be transformed into a discrete control system for testing before implementation via computer. The discrete system is to have a damped natural frequency of 2.6 rad/sec and a damping ratio of 0.707. What is the required gain and pole location?

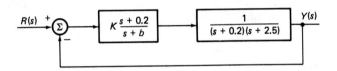

Figure 15.1   Control system for Example 15.4

SOLUTION   We will choose a sampling time of 0.1 second. Applying the ZOH process, we have

$$G(z) = \frac{0.004599z + 0.00416}{(z - 0.9801)(z - 0.7789)}$$

Applying the matched pole-zero technique, the discrete lead compensator is

$$D(z) = K_1 \frac{z - 0.9801}{z - B}$$

It is evident that the compensating network will cancel the slow pole in the discrete domain. The characteristic equation is given by

$$z^2 - (B + 0.7789 - 0.004599 K_1)z + 0.7789B + 0.00416K_1 = 0$$

From Problem 13.11 we have

$$2.6 = \frac{1}{0.1} \cos^{-1}\left(\frac{b}{2\sqrt{c}}\right)$$

which yields

$$\frac{b}{2\sqrt{c}} = 0.9664 \tag{1}$$

If we substitute this into equation (13.4.6), we obtain

$$\cos(-0.5 \ln c) - 0.9664 = 0$$

and, solving for c:

$$c = 0.5945$$

From the characteristic equation we have

$$0.7789B + 0.00416K_1 = 0.5945 \qquad (2)$$

If we substitute the value for c into equation (1), we have

$$\frac{b}{2\sqrt{0.5945}} = 0.9664$$

and, solving for b:

$$b = 1.4903$$

From the characteristic equation we have

$$B - 0.004599K_1 = 0.7114 \qquad (3)$$

Solving equations (2) and (3), we obtain $K_1 = 5.22$ and $B = 0.7354$. Therefore the required discrete compensator is given by

$$D(z) = 5.22 \frac{z - 0.9801}{z - 0.7354}$$

and the difference equation for the computer algorithm is

$$u_k = 0.7354u_{k-1} + 5.22e_k - 5.12e_{k-1}$$

The preceding example provided some insight to the design process in the discrete domain. In most cases the design engineer is more familiar with the s-domain and therefore begins his design in that domain. Once the strategy is decided upon, the compensator is transferred to the discrete domain. In our case, the matched pole-zero technique was used. In the z-domain we apply the ZOH process on the plant and complete the design to meet the specifications. We will now look at a complete design in the s-plane and its response upon transformation into the z-plane. The compensator will be the lag network.

15.5 DISCRETE LAG COMPENSATOR

The continuous form of the lag compensator is given by

$$D(s) = K \frac{s + a}{s + b}; \qquad a > b \qquad (15.5.1)$$

if we apply the matched pole-zero technique, we have

$$D(z) = K_1 \frac{z - A}{z - B}; \quad A < B \tag{15.5.2}$$

where A and B are defined as equations (15.4.3) and (15.4.4), respectively. The governing difference equation is given by equation (15.4.5). The reader should appreciate the versatility of the digital implementation of the lead and lag compensators. A minor change in coefficients provides a different compensator. Once the designer appreciates the z-domain, the design process can be carried out in the z-domain. We will investigate this at a later date, for now we will design a lag compensator in the s-domain, transfer to the z-domain, and analyze the response of the digital compensator.

EXAMPLE 15.5

The first-order plant given below is to be controlled proportionally. The design specifications require a maximum error of 10% with a 0.707 damping ratio. The design engineer decides on a lag compensator. Complete the design in the s-plane. Transfer the lag compensator using the bilinear transformation. Apply ZOH on the plant and find the damping ratio of the digital system. What is the damped natural frequency of the digital system? How does it compare with the continuous system?

$$G(s) = \frac{1}{s + 2}$$

SOLUTION   The control system is given in Fig. 15.2. We have chosen the zero location to be at $s = -5$. The additional pole from the compensator will be close to the origin, and this location of the zero will provide practical implementation as outlined in Chapter 8. The reader is encouraged to generate the root locus on the s-plane and verify the choice.

The closed-loop transfer function is given by

$$P(s) = \frac{K(s + 5)}{s^2 + s(2 + b + K) + 2b + 5K}$$

If the system is to have 10% steady-state error, the output should attain 90% of its setpoint at steady state. By applying the final-value theorem, we have

$$\frac{5K}{2b + 5K} = 0.9$$

which yields

$$K = 3.6b \tag{1}$$

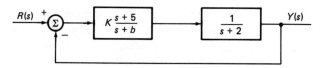

Figure 15.2   Control system for Example 15.5

If we substitute equation (1) into the characteristic equation, we obtain

$$s^2 + s(2 + 4.6b) + 20b = 0 \qquad (2)$$

The system is to exhibit a damping ratio of 0.707, therefore

$$2(0.707)\omega_m = 2 + 4.6b \qquad (3)$$

where

$$\omega_m = \sqrt{20b} \qquad (4)$$

Using equations (3) and (4), we obtain $K = 2.8$, $K = 0.875$ and $b = 0.778$, $b = 0.243$, respectively. We have decide which gain-pole combination we will use. In Chapter 8 we indicated a factor of 10 between the location of pole and zero. This guaranteed practical implementation. Therefore we will choose $K = 2.8$ and $b = 0.778$. Aside from practicality, this combination provides a higher undamped natural frequency and will increase the speed of our system for the specified damping ratio.

The design is complete, and now we will transfer it to the z-domain. The ZOH process for the plant is given by

$$G(z) = \frac{0.0906}{z - 0.819}$$

at a sampling time of 0.1 second. This was generated in Example 13.1. If we apply the bilinear transformation to the compensating network ($T = 0.1$), we have

$$D(z) = 2.8 \, \frac{20\left(\dfrac{z-1}{z+1}\right) + 5}{20\left(\dfrac{z-1}{z+1}\right) + 0.778}$$

which simplifies to

$$D(z) = 3.369 \, \frac{z - 0.6}{z - 0.925}$$

The discrete characteristic equation is given by

$$z^2 - 1.434z + 0.5745 = 0$$

From Problem 13.13 we find the damping ratio for our digital system to be 0.643. Using Problem 13.11, we find our digital system to have a damped natural frequency of 3.3 rad/sec. Our continuous system had a damped natural frequency of 2.79 rad/sec.

The damping ratio is lower for the digital system. This is expected, owing to the ZOH process. The error in the digital system is 13% by applying the final-value theorem. If the digital system is to meet the original specifications, then the gain must be reduced.

This is the design process for a complete s- to z-domain transfer. Example 15.4 seems to be the better process so far. What about a complete design in the z-plane? Let's investigate that possibility, using a lead-lag compensator on this plant.

15.6 DISCRETE LEAD-LAG COMPENSATOR

The continuous form of the lead-lag compensator is

$$D(s) = K \frac{(s + a_1)(s + a_2)}{(s + b_1)(s + b_2)}; \qquad a_1 < b_1, a_2 > b_2 \qquad (15.6.1)$$

if we apply the matched pole-zero technique, we have

$$D(z) = K_1 \frac{(z - A_1)(z - A_2)}{(z - B_1)(z - B_2)}; \qquad A_1 > B_1, A_2 < B_2 \qquad (15.6.2)$$

where

$$A_1 = e^{-a_1 T} \qquad (15.6.3)$$

$$A_2 = e^{-a_2 T} \qquad (15.6.4)$$

$$B_1 = e^{-b_1 T} \qquad (15.6.5)$$

$$B_2 = e^{-b_2 T} \qquad (15.6.6)$$

The governing difference equation is given by

$$u_k = (B_1 + B_2)u_{k-1} - B_1 B_2 u_{k-2} + K_1[e_k - (A_1 + A_2)e_{k-1} + A_1 A_2 e_{k-2}]$$

$$(15.6.7)$$

and the value of K_1 will be generated relative to system design specifications.

EXAMPLE 15.6

The control system in Example 15.5 is required to have 0% error to a step input. The damped natural frequency is to remain the same, as is the damping ratio. The error implies an integrating pole at $B_2 = 1$. This lag compensation will decrease the system speed, and we will also need some lead compensation to return the same. Therefore a lead-lag compensator is implied. Figure 15.3 provides the control scheme. Decide on the balance of the parameters to satisfy the specifications.

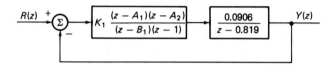

Figure 15.3 Control system for Example 15.6

SOLUTION   We will remain in the z-domain at all times. We will choose the lag-compensator zero to be equal to the pole location of the plant, $A_2 = 0.819$. This maintains our design criteria of Chapter 8, where the zero is to the left of at least two poles. It should be very close to the second pole, as we require the root locus to be pulled to the left as much as possible. For this reason the zero of the lead compensator will be chosen very close to the other zero, $A_1 = 0.8$. By moving the lead pole toward the origin of the z-plane, the speed of the system should be returned to the given specifications. The resulting compensating network is

$$D(z) = K_1 \frac{(z - 0.8)(z - 0.819)}{(z - B_1)(z - 1)}$$

The characteristic equation is given by

$$z^2 - z(B_1 + 1 - 0.0906K_1) + B_1 - 0.07248K_1 = 0$$

From Example 15.5, the designed system in the s-domain had a damped natural frequency of 2.79 rad/sec. From Problem 13.11, we have

$$2.79 = \frac{1}{0.1} \cos^{-1}\left(\frac{b}{2\sqrt{c}}\right)$$

which yields

$$\frac{b}{2\sqrt{c}} = 0.9613 \tag{1}$$

If we substitute this into equation (13.4.6), the damping ratio being 0.707, we obtain

$$\cos(-0.5 \ln c) - 0.9613 = 0$$

and, solving for c:

$$c = 0.5722$$

From the characteristic equation we have

$$B_1 - 0.07248K_1 = 0.5722 \tag{2}$$

If we substitute the value for c into equation (1), we have

$$\frac{b}{2\sqrt{0.5722}} = 0.9613$$

and, solving for b:

$$b = 1.4544$$

From the characteristic equation we have

$$B_1 = 0.0906K_1 = 0.4544 \tag{3}$$

Solving equations (2) and (3), we obtain $K_1 = 6.5$ and $B_1 = 1.043$.

There is something wrong! We have just designed a compensation network that has a pole outside the unit circle. The system will definitely be unstable. We have followed all the practices outlined in Chapter 8. There lies the problem. In Chapter 8 we specified the location of zeroes to produce a certain root locus. In Chapter 12 we indicated that zeroes do not necessarily follow the s-to-z transfer property ($z = e^{sT}$). This implies that we can place poles on the z-plane and calculate the required location of the zeroes. Zeroes can be safely specified if they are to cancel plant poles; otherwise they should be calculated. The experienced designer will be flexible in interpreting the meaning of zeroes on the z plane, while the beginner will revisit the s plane. Let's redo the problem.

Our new compensating network should be

$$D(z) = K_1 \frac{(z - A_1)(z - 0.819)}{(z - 0.2)(z - 1)}$$

We have left the location of the zero as a variable. The remaining pole was placed at $B_1 = 0.2$. This will offset the reduction of speed introduced by the integrating pole. The characteristic equation is given by

$$z^2 - z(1.2 - 0.0906K_1) + 0.2 - 0.0906A_1K_1 = 0$$

If we repeat the process with this characteristic equation, we will obtain

$K_1 = 2.52$ and $A_1 = -0.56$ for a damped natural frequency of 5.58 rad/sec. The governing difference equation is

$$u_k = 1.2u_{k-1} - 0.2u_{k-2} + 2.52e_k - 0.653e_{k-1} - 1.156e_{k-2}$$

The zero location is on the negative side of the z plane. This could have been predicted from our examples with proportional control of a second-order system. The increase in the damped natural frequency was chosen to show the benefits of the lead-lag compensation. The reader is asked to simulate this system and see if the specifications have been met. The following difference equation should be used for the plant.

$$u_k = 0.819u_{k-1} + 0.0906e_k$$

This modification was implied in Problem 14.4. If the reader refers back to Chapter 13, it will be evident that the ZOH provides a delay of one sample at the beginning. This injects a dead time or transport lag into our system. Did you think of this in Problem 14.4? We are using the ZOH process for simulating our plants and therefore have an inherent transport lag of one sample. The result is a lower damping ratio. The digital control system will reduce the damping ratio as we have indicated; the ZOH process amplifies the effect. This was required in the past chapters to enhance the theory. At this point we would like our designs to operate as specified. That is why we address the problem at this time. To achieve a more accurate representation of the plant, it is advisable to factor one time delay (z^{-1}) out of the numerator of $G(z)$ if using ZOH. Simply discard the factor. This can be thought of as an advance in ZOH process.

Let's look at transport lag.

15.7 DISCRETE TRANSPORT LAG

As stated previously, we have been dealing with transport lag throughout the simulation process. It provided an amplification of the problems with digitizing the compensating networks and in that respect was useful. In real systems it is a major problem.

For simplicity, we will assume that the dead time is equal to an integer number of sampling intervals. That is,

$$T_d = n \cdot T \tag{15.7.1}$$

This translates to a transfer block as shown in Fig. 15.4. The effects of the transport lag in the digital system are the same as indicated in the continuous system. The system phase is altered, and the system will tend to instability.

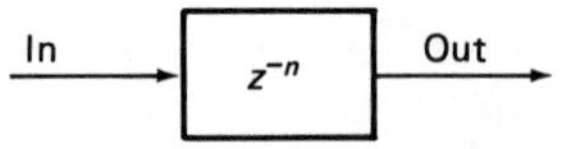

Figure 15.4 Transfer block for transport lag

From this we can understand the implications of transport lag, and the problems at the end of the chapter will provide the visual effect.

This concludes our discussion of the process of digital control. We now turn our attention to state-variable analysis in the discrete domain. If the reader is unsure of any section in Part B of this text, it would be advisable to revisit the same.

15.8 SUMMARY

In this chapter we introduced two alternative methods for transforming continuous transfer functions into discrete transfer functions: the matched pole-zero and bilinear transformation techniques. It was found that the bilinear transformation warped the frequency response. This could be alleviated by prewarping the continuous transfer function.

We investigated the discrete versions of the lead, lag, and lead-lag compensation networks. It was found that the s-plane design could be transferred with a decrease in loop gain. The reader was cautioned on zero placement in the z-plane. This warning was enhanced with an example that designed an unstable compensating network. The idea of transport lag was mentioned and the inherent transport lag of the ZOH process was identified.

Problems

15.1. If the matched pole-zero technique were applied to a bandpass filter, at what frequency would the gain be matched?

15.2. Consider the general form of the bandpass filter. If $\zeta_n = 0.1$, $\omega_n = 1$ rad/sec, and $T = 1$ second, find the equivalent discrete filter using
   (a) Approximation to differential equation.
   (b) ZOH process.
   (c) Matched pole-zero technique.
   (d) Bilinear transformation.
   (e) Bilinear transformation with prewarping.
   (f) Continuous filter.

$$D(s) = \frac{2\zeta_n \omega_n s}{s^2 + 2\zeta_n \omega_n s + \omega_n^2}$$

15.3. Plot the frequency response for each filter generated in Problem 15.2. Which represents the best approximation to the continuous filter? Is aliasing still a problem?

15.4. Plot the pole locations for all the discrete filters generated in Problem 15.2. Which filter has poles identical to the continuous poles transferred into the z plane?

15.5. Consider the critical frequency of 1 rad/sec. Generate a plot of the warped frequency ω_{PW} with respect to sampling time T. Is it necessary to use prewarping if a large sampling frequency is used?

15.6. The following represents a prewarped transfer function. What is the original transfer function if $T = 0.6$ second?

$$D_{PW}(s) = \frac{0.45609s}{s^2 + 0.45609s + 5.2}$$

15.7. From the discrete closed-loop transfer function of Problem 14.5, generate the frequency response. Generate a computer program to calculate the magnitude and phase. Plot both with respect to frequency. This becomes the closed-loop frequency response. What are the gain and the phase margin?

15.8. Repeat Problem 15.7 with the discrete closed-loop transfer function of Example 15.5.

15.9. Repeat Problem 15.7 with the final discrete closed-loop transfer function of Example 15.6.

15.10. Provide a computer simulation for the discrete control system designed in Example 15.4. Remove the transport lag of the plant due to ZOH. Do not cancel the pole and zero before simulation, as the simulation requires the full compensation network and plant present. How does the response compare with the design specifications?

15.11. Repeat Problem 15.10 for the discrete control system of Example 15.5.

15.12. Repeat Problem 15.10 for the final discrete control system of Example 15.6.

15.13. From Problem 15.12, make the location of the designed zero location a variable in your program. Investigate the response of the system with the zero moving from the origin to the outside of the unit circle on the negative real axis. Comment on the responses.

15.14. Show that the discrete control system in Example 15.6 has a minimum damped natural frequency. What is it? How is it affected if the pole at 0.2 is moved toward 0.6?

15.15. Redesign the control system in Example 15.5 using a lead-lag compensator. The system is to have 0% error at steady state for a unit step input. Use the pole-zero cancellation technique for both plant poles. The response is to exhibit a 0.707 damping ratio and is to settle within 1.5 seconds. Provide a computer simulation of your design to verify its functionality.

15.16. Consider the continuous control system given in Fig. P15.16. For the continuous system:

(a) Calculate K, damped natural frequency with the damping ratio equal to 0.707.

(b) Convert $D(s)$ to $D(z)$ using the bilinear transformation with sampling time of your choice.

(c) Convert $G(s)$ to $G(z)$ using the ZOH process with the sampling time of part (b).

(d) Using $D(z)$, $G(z)$ calculated in (b) and (c), find the damping ratio, damped natural frequency of the digital control system. If it is to meet the specifications of the continuous system, what is the required gain in the loop? Verify your calculation with a computer simulation.

(e) If the system in (d) is to have the same damped natural frequency and damping ratio as in (a), find the new digital compensator. You can leave the zero location the same and make the pole location variable.

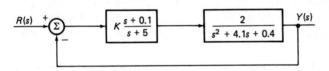

Figure P15.6   Control system for Problem 15.16

15.17. Repeat Problem 15.16 for the control system given in Fig. P15.17. Provide a root locus for this system if the gain on the compensating network is variable. What is the maximum gain? What is the gain margin at the operating point?

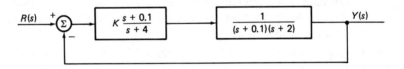

Figure P15.17   Control system for Problem 15.17

15.18. Given the discrete control system in Fig. P15.18. Find the value of the gain K_1 if the system is to exhibit a damping ratio of 0.707.

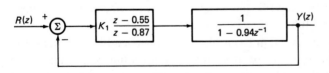

Figure P15.18   Control system for Problem 15.18

15.19. Given the control system in Fig. P15.19. Design the system to have 20% error to a unit step input. It is to exhibit a damping ratio of 0.7. Use

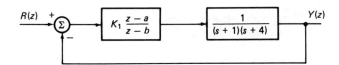

Figure P15.19   Control system for Problem 15.19

sampling time of your choice and the ZOH process for generating $G(z)$. Verify your design with a computer simulation.

15.20. Refer to Problem 8.20.   Provide a complete computer simulation for the tank level control system.   Allow for the transport lag with the transfer block given in Fig. 15.4.   Use a sampling time of 0.1 second.   Adjust the transport lag from 0 to 1.1 second.   Comment on the response.

15.21. Refer to Problem 8.22.   Provide a complete computer simulation for the position control system.   Make the derivative feedback gain variable.   Is it operating as expected?

15.22. Generate the algorithms for the lead, lag, and lead-lag compensators for your favorite microprocessor.   What are the expected computation times? What effect will the finite wordlength have on the compensators?

15.23. Refer to Problem 15.21.   Modify your program such that transport lag can be added after the control action.   Investigate the response of the position control system with various delays.

15.24. Refer to Problem 6.5.   You are to design a discrete lead-lag compensator for the speed-control system.   The steady-state error is to be 0% and the damping ratio is to be 0.707.   The damped natural frequency is to be two times the value obtained in Problem 6.5.   Make sure you choose the proper sampling time.   Verify your design with a computer simulation.

15.25. Show that the compensator $D(s)$,

$$D(s) = K \frac{s + a}{s + b}$$

can be shown as

$$D(z) = K \frac{[z - (1 + \frac{a}{b}(e^{-bT} - 1))]}{z - e^{-bT}}$$

using the ZOH approximation.   If $a \gg b$, where will the zero move on the z plane?   How does this compare with equation (15.5.2)?   What is the significance of this result, and which equation is more accurate?

Part II
Discrete

16

Discrete State-Variable Analysis

16.1 INTRODUCTION

In Chapter 9 we introduced the concept of state-variable analysis. This analysis provided a system of first-order, constant-coefficient differential equations. These could be shown in compact form via state-vector equations. The solution to these equations could be found in time by numerical methods. The solution was also possible through our classical frequency methods. It was possible to find the state transfer functions for the system and the important characteristic equation. The stability could be predicted from our standard frequency procedures. Two new concepts were introduced that provide a basis for design using state variables: controllability and observability. The first concept looked at the possibility of a state variable having no physical input. We saw that we had no control over this variable if it became unstable. The second concept looked at whether the state variable could be measured at the output. If it could not, then we could not indicate the performance of that state variable. These concepts evolved from what is considered to be modern control theory.

In this chapter we extend this analysis to the discrete domain. We will generate the system of difference equations and the state-vector equation in the discrete sense.   We will use the z-transform to find the system of transfer functions and the characteristic equation.   The ideas of controllability and observability will be investigated.   The computer simulation of the state-vector difference equation will be implemented as a form of control-system synthesis.

Let's begin with the solution to the discrete vector difference equation.

16.2 SOLUTION TO THE DISCRETE-STATE VECTOR EQUATION

Refer to Section 9.4.   If initial conditions were considered, equation (9.4.6) would be

$$\mathbf{X}(s) = [s\mathbf{I} - \mathbf{A}]^{-1}\mathbf{x}(0) + [s\mathbf{I} - \mathbf{A}]^{-1}\mathbf{BR}(s) \tag{16.2.1}$$

and the time solution would be

$$\mathbf{x}(t) = \mathbf{\Phi}(t)\mathbf{x}(0) + \int_0^t \mathbf{\Phi}(t - \tau) \cdot \mathbf{Br}(\tau)\, d\tau \tag{16.2.2}$$

where $\mathbf{\Phi}(t)$ is the state transition matrix and is defined by equation (9.4.8). Consider Fig. 16.1(a).   If we require the output at time t_2, then equation (16.2.2) becomes

$$\mathbf{x}(t_2) = \mathbf{\Phi}(t_2 - t_1)\mathbf{x}(t_1) + \int_{t_1}^{t_2} \mathbf{\Phi}(t_2 - \tau) \cdot \mathbf{Br}(\tau)\, d\tau \tag{16.2.3}$$

as t_1 represents the time for initial conditions.   If we apply ZOH, Fig. 16.1(b) represents the discrete input to our system.   Therefore equation (16.2.3) becomes

$$\mathbf{x}(kT) = \mathbf{\Phi}(T)\mathbf{x}[(k - 1)T] + \int_{(k-1)T}^{kT} \mathbf{\Phi}(kT - \tau) \cdot \mathbf{Br}[(k - 1)T]\, d\tau \tag{16.2.4}$$

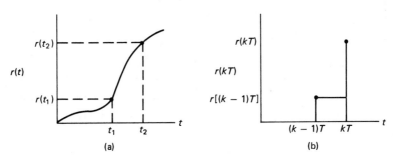

Figure 16.1   System input: (a) continuous, (b) with ZOH

which simplifies to

$$\mathbf{x}_k = \mathbf{\Phi}(T)\mathbf{x}_{k-1} + \int_{(k-1)T}^{kT} \mathbf{\Phi}(kT - \tau) \cdot \mathbf{B}\, d\tau \cdot \mathbf{r}_{k-1} \qquad (16.2.5)$$

upon noting that $\mathbf{r}[(k - 1)T]$ is constant. If we define a constant (λ) such that

$$\lambda = kT - \tau \qquad (16.2.6)$$

then equation (16.2.5) can be written as

$$\mathbf{x}_k = \mathbf{\Phi}(T)\mathbf{x}_{k-1} + \int_{0}^{T} \mathbf{\Phi}(\lambda) \cdot \mathbf{B}\, d\lambda \cdot \mathbf{r}_{k-1} \qquad (16.2.7)$$

From equation (16.2.7), we have the discrete state-vector equation

$$\mathbf{x}_k = \mathbf{A}_D\mathbf{x}_{k-1} + \mathbf{B}_D\mathbf{r}_{k-1} \qquad (16.2.8)$$

where

$$\mathbf{A}_D = \mathbf{\Phi}(T) \qquad (16.2.9)$$

$$\mathbf{B}_D = \int_{0}^{T} \mathbf{\Phi}(\lambda) \cdot \mathbf{B}\, d\lambda \qquad (16.2.10)$$

Equations (16.2.9) and (16.2.10) provide a method of converting our continuous state equations to the discrete domain. The output equation would be given by the discrete version of equation (9.3.6), which is

$$\mathbf{y}_{k-1} = \mathbf{C}\mathbf{x}_{k-1} \qquad (16.2.11)$$

EXAMPLE 16.1

Find the discrete state equations for the system given in Example 9.3. The sampling time T is to be 0.1 second.

SOLUTION   We have to find the $\mathbf{A}_D$ and $\mathbf{B}_D$ matrices. From Example 9.3 the state transition matrix is

$$\mathbf{\Phi}(t) = \begin{bmatrix} (-e^{-t} + 2e^{-2t}) & (e^{-t} - e^{-2t}) \\ (-2e^{-t} + 2e^{-2t}) & (2e^{-t} - e^{-2t}) \end{bmatrix}$$

Therefore the $\mathbf{A}_D$ matrix is given by

$$\mathbf{A}_D = \mathbf{\Phi}(0.1) = \begin{bmatrix} (-e^{-0.1} + 2e^{-0.2}) & (e^{-0.1} - e^{-0.2}) \\ (-2e^{-0.1} + 2e^{-0.2}) & (2e^{-0.1} - e^{-0.2}) \end{bmatrix}$$

which simplifies to

$$\mathbf{A}_D = \begin{bmatrix} 0.7326 & 0.08611 \\ -0.1722 & 0.9909 \end{bmatrix}$$

The $\mathbf{B}_D$ matrix is given by

$$\mathbf{B}_D = \begin{bmatrix} \int_0^{0.1} (-e^{-\lambda} + 2e^{-2\lambda})\, d\lambda & \int_0^{0.1} (e^{-\lambda} - e^{-2\lambda})\, d\lambda \\ \int_0^{0.1} (-2e^{-\lambda} + 2e^{-2\lambda})\, d\lambda & \int_0^{0.1} (2e^{-\lambda} - e^{-2\lambda})\, d\lambda \end{bmatrix} \begin{bmatrix} 4 \\ -5 \end{bmatrix}$$

relative to a unit step input. Upon integration and simplification, we obtain

$$\mathbf{B}_D = \begin{bmatrix} 0.3218 \\ -0.5347 \end{bmatrix}$$

The discrete state-vector equation is

$$\begin{bmatrix} x_1(k) \\ x_2(k) \end{bmatrix} = \begin{bmatrix} 0.7326 & 0.08611 \\ -0.1722 & 0.9909 \end{bmatrix} \begin{bmatrix} x_1(k-1) \\ x_2(k-1) \end{bmatrix} + \begin{bmatrix} 0.3218 \\ -0.5347 \end{bmatrix} r_{k-1}$$

with the system of difference equations being

$$x_1(k) = 0.7326x_1(k-1) + 0.08611x_2(k-1) + 0.3218r_{k-1}$$

$$x_2(k) = -0.1722x_1(k-1) + 0.9909x_2(k-1) - 0.5347r_{k-1}$$

The discrete output equation is given by

$$y_{k-1} = [1 - 1]\begin{bmatrix} x_1(k-1) \\ x_2(k-1) \end{bmatrix}$$

and the output difference equation is

$$y_{k-1} = x_1(k-1) - x_2(k-1)$$

The above equations can be simulated via computer, as we will see.

We turn now to the derivation of the discrete transfer function.

16.3 DISCRETE TRANSFER FUNCTION IN STATE SPACE

If we take z-transforms of equation (16.2.8), we have

$$\mathbf{X}(z) = \mathbf{A}_D z^{-1}\mathbf{X}(z) + \mathbf{B}_D z^{-1}\mathbf{R}(z) \qquad (16.3.1)$$

Multiplying both sides by z, we obtain

$$z\mathbf{X}(z) = \mathbf{A}_D\mathbf{X}(z) + \mathbf{B}_D\mathbf{R}(z) \qquad (16.3.2)$$

and, solving for $X(z)$, we have

$$\mathbf{X}(z) = [z\mathbf{I} - \mathbf{A}_D]^{-1} \cdot \mathbf{B}_D \mathbf{R}(z) \qquad (16.3.3)$$

If we take z-transforms of equation (16.2.11), we have

$$z^{-1}\mathbf{Y}(z) = z^{-1}\mathbf{C}\mathbf{X}(z) \qquad (16.3.4)$$

and upon simplification, we obtain

$$\mathbf{Y}(z) = \mathbf{C}\mathbf{X}(z) \qquad (16.3.5)$$

If we substitute equation (16.3.3) into (16.3.5), we have

$$\mathbf{Y}(z) = \mathbf{C} \cdot [z\mathbf{I} - \mathbf{A}_D]^{-1} \cdot \mathbf{B}_D \mathbf{R}(z) \qquad (16.3.6)$$

Equation (16.3.6) represents an output-to-input relationship. Therefore the expression

$$\mathbf{P}(z) = \mathbf{C}[z\mathbf{I} - \mathbf{A}_D]^{-1} \cdot \mathbf{B}_D \qquad (16.3.7)$$

represents the discrete transfer-function matrix.

The transfer function is an $m \times i$ matrix with the following entries:

$$\mathbf{P}(z) = \begin{bmatrix} P_{11}(z) & P_{12}(z) & \cdots & P_{1i}(z) \\ \cdot & & & \\ \cdot & & & \\ \cdot & & & \\ P_{m1}(z) & P_{m2}(z) & \cdots & P_{mi}(z) \end{bmatrix} \qquad (16.3.8)$$

where

$$P_{mi}(z) = \frac{Y_m(z)}{R_i(z)} \qquad (16.3.9)$$

From equation (16.3.7) it is evident that the denominator of the transfer-function matrix is given by its determinant. Therefore our discrete characteristic equation is

$$\det [z\mathbf{I} - \mathbf{A}_D] = 0 \qquad (16.3.10)$$

The reader should appreciate the similarity of these results as compared to the continuous system.

Let's find the transfer-function matrix for the system in Example 16.1.

EXAMPLE 16.2

For the discrete state system in Example 16.1, find the discrete transfer-function matrix. Find the time solution to a unit step input and the location of the poles on the z-plane. Compare these results to those obtained in Example 9.4.

SOLUTION   If we begin with $[z\mathbf{I} - \mathbf{A}_D]$, we have

$$[z\mathbf{I} - \mathbf{A}_D] = \begin{bmatrix} z & 0 \\ 0 & z \end{bmatrix} - \begin{bmatrix} 0.7326 & 0.08611 \\ -0.1722 & 0.9909 \end{bmatrix}$$

$$= \begin{bmatrix} z - 0.7326 & -0.08611 \\ 0.1722 & z - 0.9909 \end{bmatrix}$$

The inverse matrix is

$$[z\mathbf{I} - \mathbf{A}_D]^{-1} = \frac{1}{z^2 - 1.72352 + 0.7408} \begin{bmatrix} z - 0.9909 & 0.08611 \\ -0.1722 & z - 0.7326 \end{bmatrix}$$

From equation (16.3.7), the transfer-function matrix is

$$\mathbf{P}(z) = \left(\frac{1}{z^2 - 1.7235z + 0.7408} \right) \begin{bmatrix} 1 & -1 \end{bmatrix} \begin{bmatrix} z - 0.9909 & 0.08611 \\ -0.1722 & z - 0.7326 \end{bmatrix} \begin{bmatrix} 0.3218 \\ -0.5347 \end{bmatrix}$$

which simplifies to

$$P(z) = \frac{0.8565}{z - 0.9048}$$

If the input is a unit step, the output is

$$Y(z) = \frac{0.8565}{(1 - z^{-1})(z - 0.9048)}$$

and the output difference equation would be

$$y_k = 9 - 9\,(0.9048)^{k-1}$$

after taking inverse z-transforms.   The reader is encouraged to tabulate the values of the difference equation and compare them to the output obtained in example (9.4).   Remember that each sample is taken at 0.1 second intervals.

   The characteristic equation is given by

$$z^2 - 1.7235z + 0.7408 = 0$$

and the pole locations are at $z = 0.9048$ and $z = 0.8187$.   This corresponds to the mapping of the s-plane poles to the z-plane.   If we are to simulate this system in real time, a computer program is required.

   Let's develop the computer algorithm to solve the discrete system of difference equations.

16.4 COMPUTER SIMULATION OF A DISCRETE STATE SYSTEM

The simulation of the discrete state system requires n stacks that have 2 locations. Recall that the previous simulation of an nth-order system required 2 stacks with n locations. Figure 16.2 provides the stack orientation for our

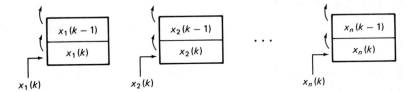

Figure 16.2   Stack representation of discrete state equation

discrete state system. It is evident that we require one shift process for each stack. In a high-level language the stacks can be represented by a 2×1 array. If we use the BASIC language, we can implement this stack shifting as indicated in Fig. 16.3.

```
200   X1(1) = X1(0)
210   X2(1) = X2(0)
        .
        .
        .
***   Xn(1) = Xn(0)
```

Figure 16.3   BASIC program to implement stack shifting

Once the stacks are shifted, the difference equations can be implemented. The program in Fig. 16.4 provides the BASIC implementation of the discrete state equations. The choice of variables should be self-explanatory.

```
1000 X1(0) = A11*X1(1)+...+A1n*Xn(1) + B11*R1(1)+...+B1i*Ri(1)
1010 X2(0) = A21*X1(1)+...+A2n*Xn(1) + B21*R1(1)+...+B2i*Ri(1)
        .
        .
        .
**** Xn(0) = An1*X1(1)+...+Ann*Xn(1) + Bn1*R1(1)+...+Bni*Ri(1)
```

Figure 16.4   BASIC implementation of discrete state equation

To implement the discrete state output equation, the BASIC program in Fig. 16.5 can be used.

```
2000    Y1(1)  =  C11*X1(1)  +...+  C1n*Xn(1)
2010    Y2(1)  =  C21*X1(1)  +...+  C2n*Xn(1)
                      .
                      .
                      .
******Ym(1)  =  Cm1*X1(1)  +...+  Cmn*Xn(1)
```

Figure 16.5   BASIC implementation of discrete state output equation

If we combine the programs in Figs. 16.3, 16.4, and 16.5, we will generate a general program for the solution of an nth-order discrete state system. The details missing are better handled with an example. Let's simulate the system in Example 16.1.

EXAMPLE 16.3

Provide a computer simulation for the discrete system given in Example 16.1. Run the program and compare to the continuous time output.

SOLUTION   The sampling time of this system was 0.1 second. If we inspect equations (16.2.8) and (16.2.11), we also require a stack for the inputs and outputs. The complete BASIC program is provided in Fig. 16.6.

```
10      FOR SAMPLE% = 0 TO 100
20          *** REM STEP INPUT
30          R1(0) = 1
200         X1(1) = X1(0)
210         X2(1) = X2(0)
220         *** REM SHIFT INPUTS AND OUTPUTS
230         R1(1) = R1(0)
240         Y1(1) = Y1(0)
1000        X1(0) = 0.7326*X1(1) + 0.08611*X2(1) + 0.3218*R1(1)
1010        X2(0) = -0.1722*X1(1) + 0.9909*X2(1) - 0.5347*R1(1)
2000        Y1(1) = X1(1) - X2(1)
3000        PRINT "Time =";SAMPLE%*0.1,Y1(1)
9999    NEXT SAMPLE%
```

Figure 16.6   BASIC program for simulation of system in Example 16.1

Table 16.1 compares the simulation output with the continuous system response. It is evident that our discrete state system is operating as expected.

TABLE 16.1 Comparison of
Simulation to Continuous Output

Time (sec)	Simulation	Continuous
0.0	0.0000	0.0000
0.1	0.8565	0.8565
0.2	1.6315	1.6314
0.3	2.3326	2.3326
0.4	2.9670	2.9671
0.5	3.5411	3.5412
1.0	5.6883	5.6891
2.0	7.7797	7.7820
3.0	8.5486	8.5519
4.0	8.8313	8.8352
5.0	8.9352	8.9394
10.0	8.9952	8.9996

The reader is encouraged to modify the program so that the state variables are also tabulated. Then a graphing of the same would provide information on their relevance to the system output.

To complete our analysis of the discrete state-variable system we still have to investigate the existence of controllability and observability. Let's turn our attention to this area.

16.5 CONTROLLABILITY OF DISCRETE FUNCTIONS

The definition of controllability remains as outlined in Chapter 9. The mathematical testing for controllability will be given by equation (9.6.1), with the matrices being from the discrete system. Therefore the controllability matrix in the discrete state system is given by

$$\mathbf{M}_c = [\mathbf{B}_D \vdots \mathbf{A}_D\mathbf{B}_D \vdots \ldots \vdots \mathbf{A}_D^{n-1}\mathbf{B}_D] \tag{16.5.1}$$

The discrete control system is totally controllable if its controllability matrix is of full rank.

EXAMPLE 16.4

Investigate the controllability of the system given in Example 16.1.

SOLUTION   This system is second order; therefore the controllability matrix is given by

$$\mathbf{M}_c = [\mathbf{B_D} \,|\, \mathbf{A_D B_D}]$$

and upon multiplication we have

$$\mathbf{M}_c = \begin{bmatrix} 0.3218 & 0.1897 \\ -0.5347 & -0.5852 \end{bmatrix}$$

The determinant of the controllability matrix is nonzero; therefore the discrete system is totally controllable.

16.6 OBSERVABILITY OF DISCRETE FUNCTIONS

The definition of observability remains as outlined in Chapter 9. The mathematical testing for observability will be given by equation (9.6.2), with the matrices being from the discrete system. Therefore the observability matrix in the discrete state system is given by

$$\mathbf{M}_o = \begin{bmatrix} \mathbf{C} \\ \overline{} \\ \mathbf{CA_D} \\ \overline{} \\ \cdot \\ \cdot \\ \cdot \\ \mathbf{CA_D^{n-1}} \end{bmatrix} \tag{16.6.1}$$

The discrete control system is totally observable if its observability matrix is of full rank.

EXAMPLE 16.5

Investigate the observability of the system given in Example 16.1.

SOLUTION   The observability matrix for this system is given by

$$\mathbf{M}_o = \begin{bmatrix} \mathbf{C} \\ \overline{} \\ \mathbf{CA_D} \end{bmatrix}$$

and upon multiplication, we have

$$M_o = \begin{bmatrix} 1 & -1 \\ 0.9048 & 1.077 \end{bmatrix}$$

The determinant of the observability matrix is nonzero; therefore the discrete system is totally observable.

This concludes our study of discrete control systems. The reader is encouraged to pursue further any area of this vast and expanding field. The material contained herein is simply an introduction.

16.7 SUMMARY

In this chapter we investigated the discrete state control system. By applying the ZOH process on the continuous state system, we generated the new coefficient matrices for the state-vector difference equations. The discrete state-transfer function matrix was developed. The discrete characteristic equation was generated. The simulation of the state system was provided by stack representation and then implemented with the BASIC language. To fully complement Chapter 9, we concluded by discussing the controllability and observability of the discrete state system.

Problems

16.1. Refer to Problem 9.1. Choose an acceptable sampling time and generate the discrete state equations for this system.

16.2. Refer to Problem 9.2. Choose an acceptable sampling time and generate the discrete state equations for this system. Assume $m = 1$, $B = 6$, and $K = 4$, all with appropriate units.

16.3. Refer to Problem 9.5. Choose an acceptable sampling time and generate the discrete state-transfer function. What is the damping ratio of the discrete system? How does it compare to the continuous system?

16.4. Refer to Example 9.5. If $T = 1$ second, generate the discrete transfer-function matrix from the continuous state system. Isolate the characteristic equation. Plot the poles on the z-plane. How do they compare with the poles on the s-plane transferred to the z-plane?

16.5. Refer to Problem 9.6. Choose an acceptable sampling time and generate the discrete state-transfer function.

16.6. Refer to Problem 9.7. Repeat Problem 16.5.

16.7. Refer to Problem 9.14. Choose an appropriate sampling time and repeat parts (b) to (g) in the discrete sense.

16.8. Provide a computer simulation of the system in Problem 16.7.

16.9. Consider the program in Fig. 16.6. If the plant represented by this simulation is to be proportionally controlled,

   (a) Calculate the maximum gain.

   (b) Calculate the gain required for the digital system to have a 0.707 damping ratio.

   (c) Modify the program in Fig. 16.6 to provide the computer simulation for the control system. Plot the outputs of all state variables. How does the system output compare with the design specifications? Modify lines 1000 and 1010 by replacing the term R1(1) with R1(0). Rerun the program. Is there any difference? Why?

16.10. The plant in Problem 16.7 is to be proportionally controlled. What gain is required for the embedded second-order digital system to have a 0.707 damping ratio? Modify the program from Problem 16.8 to simulate the control system. Does it meet specifications?

Greek Alphabet

Forms		Names	Forms		Names
A	α	alpha	N	ν	nu
B	β	beta	Ξ	ξ	xi
Γ	γ	gamma	O	o	omicron
Δ	δ	delta	Π	π	pi
E	ε	epsilon	P	ρ	rho
Z	ζ	zeta	Σ	σ	sigma
H	η	eta	T	τ	tau
Θ	θ	theta	Υ	υ	upsilon
I	ι	iota	Φ	φ	phi
K	κ	kappa	X	χ	chi
Λ	λ	lambda	Ψ	ψ	psi
M	μ	mu	Ω	ω	omega

Index

T

Tach generator, 46
Transfer function:
    definition, 37–39
    properties, 39–42
Transport lag, 16
    discrete, 279–80

U

Undamped natural frequency, 76
Underdamped, 35, 78
Unity feedback, 61
Unstable, 50, 51

Z

Z-transforms:
    derivations, 195–99
    inverse, 202–5
    properties, 199–202
Zeroes, 40
Zero Order Hold (ZOH), 229–34